Fulfilling the Promise

High-school students examining a lens of a cow's eye.
Courtesy, UC San Francisco photo by David Powers.

Fulfilling the Promise

BIOLOGY EDUCATION IN THE NATION'S SCHOOLS

Committee on High-School Biology Education
Board on Biology
Commission on Life Sciences
National Research Council

NATIONAL ACADEMY PRESS
Washington, D.C.   1990

National Academy Press • 2101 Constitution Avenue, N.W. • Washington, D.C. 20418

NOTICE: The project that is the subject of this report was approved by the Governing Board of the National Research Council, whose members are drawn from the councils of the National Academy of Sciences, the National Academy of Engineering, and the Institute of Medicine. The members of the committee responsible for the report were chosen for their special competences and with regard for appropriate balance.

This report has been reviewed by a group other than the authors according to procedures approved by a Report Review Committee consisting of members of the National Academy of Sciences, the National Academy of Engineering, and the Institute of Medicine.

This Board on Biology study was supported by the Howard Hughes Medical Institute.

Any opinions, findings, conclusions, or recommendations expressed in this publication are those of the authors and do not necessarily reflect the views of the Howard Hughes Medical Institute.

Library of Congress Cataloging-in-Publication Data

National Research Council (U.S.). Committee on High-School Biology Education.
    Fulfilling the promise : biology education in the nation's schools
/ Committee on High-School Biology Education, Board on Biology, Commission on
Life Sciences, National Research Council.
      p.   cm.
    Includes bibliographical references and index.
    ISBN 0-309-04243-7
    1. Biology—Study and teaching—United States. I. Title.
QH319.A1N38   1990   90-42248
574'.071'0973—dc20   CIP

Printed in the United States of America

COMMITTEE ON HIGH-SCHOOL BIOLOGY EDUCATION

TIMOTHY H. GOLDSMITH (*Chairman*), Yale University, New Haven, Connecticut
CLIFTON POODRY (*Vice Chairman*), University of California, Santa Cruz

R. STEPHEN BERRY, University of Chicago, Chicago, Illinois
RALPH E. CHRISTOFFERSEN, Smith Kline and French Laboratories, King of Prussia, Pennsylvania
JANE BUTLER KAHLE, Miami University, Oxford, Ohio
MARC W. KIRSCHNER, University of California, San Francisco
JOHN A. MOORE, University of California, Riverside
DONNA OLIVER, Bennett College, Greensboro, North Carolina
JONATHAN PIEL, *Scientific American*, New York, New York
JAMES T. ROBINSON, Boulder, Colorado
JANE SISK, Calloway County High School, Murray, Kentucky
WILMA TONEY, Manchester High School, Manchester, Connecticut
DANIEL B. WALKER, San Jose State University, San Jose, California

Special Advisors

PAUL DeHART HURD, Palo Alto, California
JOHN HARTE, University of California, Berkeley

Former Members

EVELYN E. HANDLER (*Chairman, 1987-1988*), Brandeis University, Waltham, Massachusetts
MICHAEL H. ROBINSON (*1987-1988*), National Zoological Park, Washington, D.C.
MARY BUDD ROWE (*1987-1989*), University of Florida, Gainesville, Florida
DAVID T. SUZUKI (*1987-1989*), University of British Columbia, Vancouver, British Columbia, Canada

National Research Council Staff

JOHN E. BURRIS, Study Director
DONNA M. GERARDI, Staff Officer
WALTER G. ROSEN, Consultant
NORMAN GROSSBLATT, Editor
MARY KAY CERIANI, Senior Secretary

The National Academy of Sciences is a private, nonprofit, self-perpetuating society of distinguished scholars engaged in scientific and engineering research, dedicated to the furtherance of science and technology and to their use for the general welfare. Upon the authority of the charter granted to it by the Congress in 1863, the Academy has a mandate that requires it to advise the federal government on scientific and technical matters. Dr. Frank Press is president of the National Academy of Sciences.

The National Academy of Engineering was established in 1964, under the charter of the National Academy of Sciences, as a parallel organization of outstanding engineers. It is autonomous in its administration and in the selection of its members, sharing with the National Academy of Sciences the responsibility for advising the federal government. The National Academy of Engineering also sponsors engineering programs aimed at meeting national needs, encourages education and research, and recognizes the superior achievements of engineers. Dr. Robert M. White is president of the National Academy of Engineering.

The Institute of Medicine was established in 1970 by the National Academy of Sciences to secure the services of eminent members of appropriate professions in the examination of policy matters pertaining to the health of the public. The Institute acts under the responsibility given to the National Academy of Sciences by its congressional charter to be an advisor to the federal government and, upon its own initiative, to identify issues of medical care, research, and education. Dr. Samuel O. Thier is president of the Institute of Medicine.

The National Research Council was organized by the National Academy of Sciences in 1916 to associate the broad community of science and technology with the Academy's purposes of furthering knowledge and of advising the federal government. Functioning in accordance with general policies determined by the Academy, the Council has become the principal operating agency of both the National Academy of Sciences and the National Academy of Engineering in providing services to the government, the public, and the scientific and engineering communities. The Council is administered jointly by both Academies and the Institute of Medicine. Dr. Frank Press and Dr. Robert M. White are chairman and vice chairman, respectively, of the National Research Council.

Preface

The seed for this study was germinating in the Commission on Life Sciences of the National Research Council when the Board on Biology was created in 1984. The board's attention was initially drawn to high-school biology by the controversy over the inclusion of evolution in the school curriculum, but in a one-day workshop with teachers and textbook publishers it quickly became apparent that myriad other problems beset the teaching of science. As the present study got under way, it was our intention to focus on the high-school biology curriculum, but we found that restricted goal elusive. Perhaps I can explain why by paraphrasing one of our reviewers, who characterized this report as describing the "ecology" of science education. That puts it well, for this is a report about complex relationships—how failure of learning in high-school science has its origins in elementary school, how texts, tests, teacher education, colleges and universities, and political and economic assumptions all contribute to the status quo, and how difficult it is to alter any one element alone and expect any meaningful change in the entire system. There is of course a history, too—how the nation's educational system got into its present state, and why previous efforts at reform of science education have been so ephemeral. In short, as our deliberations progressed, we were compelled by the nature of the problem to broaden the scope of our analysis.

To whom is this volume addressed? The simple answer is to everyone interested in education and the performance of our schools: teachers, parents, scientists, school boards, school administrators, science educators, legislators, and all who make or support policy that affects our schools. The need for change is pervasive and will have to be accomplished on a broad front, because the very intellectual and cultural environment in which both children and their teachers are exposed to science must be altered. Consequently, there is work to be done by everyone.

As this is a report of the National Academy of Sciences/National Research Council, it is appropriate that we have a special message for scientists, particularly those who teach in colleges and universities. Traditionally aloof from the world of precollege education, our institutions of higher learning in fact contribute to the calamity. But of this, more in the report.

Several years elapsed before the National Research Council was able to find a willing sponsor for this study, and we are grateful to the Howard Hughes Medical Institute, whose officers and trustees shared some of the same concerns about the distressing state of science education and who have generously underwritten the assessment that is presented in this volume. Like all who labor on such analyses and reports, we who made up the committee obviously hope that we have made a useful contribution toward the solution of a complex national problem, but we would also like to salute the stewards of the Howard Hughes Medical Institute for enabling us to make our case. In particular, Purnell Choppin and Joseph Perpich have displayed vision and leadership in directing attention and resources to the problems of science education.

I would like to share a reminiscence about that early one-day meeting that led to this study, because it introduces a theme we have strived to develop in the book. I had not realized until that day the depth of isolation and abandonment now felt by many able and dedicated teachers who had participated in summer institutes for secondary-school science teachers first sponsored by the National Science Foundation 20-25 years ago. Those experiences created a sense of community, a feeling of belonging to a larger guild of professional scientists that was both helpful and sustaining, but that largely melted away with the ending of the federal programs in the early 1980s. Those of us who teach know well that enthusiasm is indeed infectious, and I have had great trouble reconciling the cold, analytical studies that purport to demonstrate the ineffectiveness of those summer programs on student learning—studies that were used as part of the justification for largely terminating federal involvement in science education—with the joyous memories I hear whenever I interact with teachers from that era. This is an issue discussed at greater length in our report, but it is one illustration of how fragile is the place of teachers. As a nation we ask teachers to do a job requiring dedication and professional performance, but we sabotage the professionalism of teaching in countless ways. There is much more to successful teaching than loud cries for "accountability" might have one believe, and the need to create an appropriately professional environment for teachers is at the heart of our problem.

The experience of working with this committee has been personally rewarding, for it has demonstrated how a diverse group of individuals representing practicing teachers, research scientists, science educators, university teachers, school administrators, and others can work harmoniously on a complex educational matter. We did not always start with agreement, but our areas of disagreement always shrank dramatically with discussion, and in the end, little of substance separated us. Mutual respect and a conscientious effort to address the central issue before us inevitably prevailed. The experience makes me optimistic that the approaches we have outlined in this book can in fact be successfully implemented in the larger community by forging new working alliances of concerned participants.

Finally, with the rest of the committee I would like to thank John Burris, Donna Gerardi, and Walter Rosen, whose staff work made the study go. They have organized meetings, pursued background papers, and answered countless queries in a superbly professional manner.

TIMOTHY H. GOLDSMITH, *Chairman*
Committee on High-School Biology Education

Contents

1

Introduction

The wish to educate every citizen is at the foundation of American democracy. It constitutes a goal that has taken root in many other countries, and it is one of our national contributions of which we are justifiably proud. There is currently great concern, however, about the quality and effectiveness of our public education system. That concern was focused in the 1983 report *A Nation at Risk* by the National Commission on Excellence in Education, and it has since intensified with the growing awareness that a citizenry with an understanding of the role of science and technology is the key to our nation's future economic security.

This committee was assembled by the National Research Council's Board on Biology to confront a piece of the problem—the state of the high-school biology curriculum—but we quickly recognized how interlocked are the practices that maintain the present unsatisfactory state of precollege science education. We also saw that the teaching of biology provides a paradigm, not only illustrating what is generally awry in conveying science to children, but also providing unique opportunities for improvement.

Our children's knowledge of science is often compared unfavorably with that of students in other countries (IEA, 1988; Lapointe et al., 1989), but we do not have to look over our shoulders to find cause for alarm. A recent test of biological information taken by approximately 12,000 American high-school students (Mullis and Jenkins, 1988) yielded an astonishing result: fully half the students who had not taken a course in biology did as well as or better than 40% of the students who had taken such a course. Clearly, a great many children are learning almost nothing in their biology courses.

About 75-80% of high-school students take a course in biology, usually in the ninth or tenth grade. Only about 30% of the students continue with science by enrolling in chemistry, and only half of those carry on further by studying physics in the twelfth grade (Welch et al., 1984). This precipitous decline in enrollments suggests that something is profoundly wrong in how we inspire interest in science and convey knowledge about science to the next generation. Simply mandating an additional year of science for high-school graduation will not improve science education.

Because in most schools biology occupies a pivotal place in the curriculum at the start of the high-school sequence of science courses, it is the logical course to examine first to understand our failures. Furthermore, the nature of the subject presents unique opportunities. Biology has important things to tell children about themselves and should therefore be intrinsically interesting to them. Our lack of success in adequate instruction is especially troublesome, because for most students we are failing to relate the science of life to the experience of living.

What are appropriate goals for science education in our schools? What degree of scientific literacy can we reasonably expect of most children? In general terms, science education in kindergarten through twelfth grade should enable students to:

• Apply the methods of scientific observation and evaluation in decision-making.

• Distinguish observations from inference, compare personal "theories" with scientific theories, and understand the functions of hypothesis and theory in science and how theories are developed and tested.

• Understand the limitations of small numbers of observations in generating scientific knowledge.

• Deliberate thoughtfully with peers and adults about the outcomes and meanings of investigations and about how conceptual contradictions can be resolved through reinvestigation.

In particular, biology education in kindergarten through twelfth grade should enable students to understand:

• Basic concepts of biology.

• How to lead healthy lives through a knowledge of how their bodies work and can be abused.

• The diversity, evolution, and interdependences of the biosphere and the students' role as future stewards of the environment.

• The role of biotechnology and its impact on their lives.

The committee has developed some firm conclusions about what is wrong with biology education and how it has failed to fulfill these goals. We are convinced, however, that successful efforts to improve the classroom teaching of biology must address numerous interacting forces that maintain the inadequate status quo. Nothing short of a massive attack at a variety of points will produce the desired result—a citizenry with a much firmer understanding of humankind and of the natural world in which we all live.

This report discusses the following issues:

- Exposure to science usually does not begin early enough in the schools, and it is generally of such poor quality that students learn to dislike science. In the elementary grades, there is a great need for teachers to introduce children to the active exploration of natural phenomena in a manner that does not kill future interest in science.

- The middle-school life-science course is usually a junior version of the tenth-grade biology course and shares many of its defects. Several models for its redesign are under development and should be watched closely. Given the age and interests of middle-schoolers, the new curricula with a focus on human biology seem particularly promising.

- The present high-school biology curriculum is much too inclusive, burdened with vocabulary, and short on concepts. Most students see it as boring or irrelevant.

- Present modes of instruction are mostly unsuccessful in presenting science as a process of discovery by which we learn about the world, and the use of laboratories is inadequate.

- In many classrooms, the textbook defines the curriculum, but most textbooks are poorly structured and often inaccurate or misleading. In attempting to cover large amounts of material, they are superficial and uninteresting and fail to convey an understanding of biological principles.

- Standardized testing has become the primary method of assessment, and the results are commonly and inappropriately used to gauge the performance of both programs and individual students. Moreover, because the tests emphasize name recognition, they drive some of the less desirable features of the curriculum.

- Opportunities for effective inservice and preservice preparation of teachers are seriously inadequate.

- State and local agencies, either through inaction or through educationally misguided decisions, often reinforce the errors of the present system.

- Various infrastructural elements in the educational system detract from the professionalism of teaching; these elements work against both the recruitment and retention of able teachers and the exercise of initiative and imagination in the classroom.

- The shifting demographics of the nation—by the year 2000 one-third of our schoolchildren will be members of minority groups—present teachers with intensifying pedagogical challenges. Children do not arrive in school as blank slates on which anything can be written. If they are to learn effectively, their instruction must be related to the world they know and must be free of attitudes and expectations based on ethnic or other stereotypes. Teachers will have to deal with increasing cultural, social, and economic diversity among their students, and the underrepresentation of minority groups in the teaching profession is becoming worse.

- There are no mechanisms or long-term support for developing and evaluating science curricula.

- There is a general lack of leadership in the reform of science education at all levels. For example, not only have most of our major colleges and universities failed to design effective science curricula for their nonscience students, but even more have failed to recognize the need for scientists to support precollege science education.

In the pages that follow, we consider in detail the above issues and the many interactions among them. Some of our recommendations can be implemented immediately, at least in pilot programs, by individuals and groups willing to take the initiative. Others will require the generation of consensus through concerted efforts by many individuals with vastly different responsibilities. Substantial improvement in vitalizing science education will not be achieved by tinkering with the system. It will require recognition that the problem is complicated and will require the contributions not only of teachers, but of those who teach the teachers, teachers unions, educational administrators, makers of tests, publishers of textbooks, members of school boards, parents, politicians, and scientists. Major financial commitments will have to be made by foundations, industry, and federal, state, and local governments. A great deal is at stake, and there are important roles to be played by everyone.

The moment for effective action is now. Much is being said about the inadequate state of public education, and, more important, many models for improving the teaching and learning of science are now being tried at the local level. There is thus much cause for optimism. However, we face a great national challenge that requires national leadership. In Chapter 8 we present an agenda and an opportunity for the community of scientists to participate.

2

Where Are We Now?
The Motivation for Change

EDUCATION IN SCIENCE FOR THE TWENTY-FIRST CENTURY

The growth of scientific knowledge during the twentieth century has been without precedent in human history; science and technology permeate our culture. Some degree of familiarity with how scientific knowledge is obtained, with certainty and uncertainty, with the living and nonliving world, with basic mathematical ideas (numeracy), with how an understanding of nature and of the human body contributes to healthy lives and a safer world—in short, the basic foundation that is referred to as scientific literacy—has become an educational necessity.

In an increasingly technological society competing for world markets, the need of businesses and industries for the scientifically literate will continue to expand. Population growth has placed new strains on the environment— massive pollution of air and water, deforestation and extinction of species, global warming and shifts in climate, and alterations in the ozone shield. We are engaging in the greatest uncontrolled experiment in human history, and the outcome is far from clear. Some are reassured by those economists who are confident that there will be technological fixes to such problems. Others—more taken by the briefness of human experience since the industrial revolution, the accelerating pace of change, and the ecological concept of a finite carrying capacity—are far less sure. What is certain, however, is that these issues are here to stay; and a necessary step to their resolution in a democratic society will be increasing the scientific sophistication of elected officials and the public. As we approach the twenty-first century, science must be accepted as a basic subject that must be taught in an understandable fashion to all students.

5

BIOLOGY IN ELEMENTARY SCHOOL

Learning is cumulative. Poor background in or an inappropriate attitude about a particular body of knowledge impedes or even prevents understanding. Even bright, college-educated adults often retreat to strictly rote learning of isolated facts when confronted with unfamiliar or threatening material. The progress of young learners doubtless is even more sensitive to background and attitude. Attempts by an instructor to promote learning of sophisticated concepts at a time when the students are still at an early learning state tend only to frustrate, and frustration leads to poor attitudes and phobias about the subject.

The logic of the educational system is that learning achieved during the elementary years (generally grades K-5 or grades K-6) is built on during middle school; learning achieved in middle school (generally grades 6 and 7, grades 6-8, or grades 7 and 8) is built on during high school; and so on through college. But what happens if the elementary-school years do not result in the basic learning on which achievement in middle school and high school is predicated? Can a student perform at the expected level of, say, tenth-grade biology if the previous 9 or 10 years of schooling has not provided a substantial foundation in life science? The national condition of high-school science—as mentioned in Chapter 1 and as described in more detail later in this chapter—indicates that the answer is no.

As we argue in detail later in this report, the high-school biology course should be a synthetic treatment of important concepts and of how these concepts can shape our understanding of ourselves and our planet. If, as is true today, most citizens will never take another course in biology, or indeed any science, the high-school biology course will need to leave all students with knowledge and skills to help them interpret a complex world. But few students enter the high-school course with the background and perspective needed for such a demanding outcome. Instead, they arrive with poor attitudes toward science and often a need for remedial instruction, and, as noted earlier, they leave knowing little more than when they arrived. Their previous schooling not only has devoted little time to the study of science, but has usually been misdirected toward rote learning and textbook-centered lessons. The poor degree of learning at even the strictly factual level has been documented by national and international comparisons (IEA, 1988; Mullis and Jenkins, 1988; Lapointe et al., 1989).

The message is clear: If science literacy is to be achieved, science must be allocated as much time in elementary and middle schools as has historically been allocated to reading, writing, and mathematics. In today's world, science is an educational "basic," and it must no longer be regarded as optional or as a special "enriching" experience to be presented only if time permits. Failure to educate and excite students in science in the early years is a primary reason for the inadequacy of the learning of science in later years. Thus, no reform of science education is likely to be successful until science is taught effectively in elementary school.

Which scientific topics are emphasized and when must be based on the developmental state of the learner. Matching the curriculum to the interests and abilities of the children will improve learning. Furthermore, the major objective of elementary-school science courses should be to foster positive attitudes about and respect for the natural world, rather than to acquire detail.

Children in grades 3-6 might be especially sensitive, because it is in those grades that "likes" and "dislikes" become established. Boring or nonunderstandable science classes during those years can permanently destroy an interest in science. Conversely, exciting experiences can lead to science-related career decisions.

The elementary-school years present an opportunity for teaching about the natural world that the nation's schools have failed to grasp. Much that we know about learning, however, suggests that our success in teaching science in later years will remain limited until we capture the imaginations of children in elementary school. Later in this report, we offer specific suggestions for teaching biology to children in elementary school.

BIOLOGY IN THE MIDDLE GRADES

Biology in the middle grades is titled "life science" primarily to distinguish it from biology courses in high schools. Students in life-science courses are 13 or 14 years old, typically in the early adolescent phase of biological and social development. The life-science course is likely to be their first in a single discipline and taught by someone designated a science teacher.

The Rationale for Middle Schools

A word about the history of middle schools will be helpful in putting their curricular goals in perspective. Junior high schools were founded in 1905, because of the recognition by psychologists, biologists, and educators that the biological and social factors that identify the early adolescent are sufficiently distinctive to justify a distinctive educational program. Until the middle 1950s, the life-science course drifted toward a diluted version of high-school biology or was a maze of discrete topics distributed throughout general-science textbooks. The initial educational objective of meeting the developmental needs of early adolescents vanished and was replaced by the notion that life science in junior high schools should manage to prepare students for high-school biology. Even after 50 years, a life-science curriculum thought to be essential for the education of early adolescents still had not emerged.

In the late 1950s, educators recommended a reorganization of the junior high school and the formation of middle schools, omitting the ninth grade (Conant, 1960; Van Til et al., 1961). The advocates of the middle school reasserted the belief that the curriculum of the new school organization should fit the developmental needs of the early adolescent—the same educational concept that had given rise to the junior high school (Eichhorn, 1966). Middle schools consist of various grade combinations below the ninth grade (e.g., grades 6 and 7, grades 6-8, and grades 7 and 8); the ninth grade is part of the high school.

The projects for improving science curricula supported by the National Science Foundation during the 1960s and early 1970s developed a new series of science courses that were tailored for the traditional junior high school. Those courses were designed to represent better the "structure" of particular disciplines and to foster the development of skills in scientific inquiry. But because they did not address the needs of the adolescent, they were not consistent with the rationale for the middle-school curriculum. Projects developed particularly for early adolescents that were based on modular materials and that involved students in investigating natural phenomena were unable to penetrate the market and become established as alternatives to existing junior-high-school materials (Clark, 1969).

The middle-school movement lost its identity during the 1970s as it became a forum for addressing problems of segregation, shifts in enrollment, and the use of buildings. Serious educational problems of the early adolescent were dealt with on an ad hoc basis, if at all. Moreover, the innovative science courses of the 1960s lost their significance with the insertion of minicourses focused on specific topics, such as the metric system, energy problems, space, environmental sciences, and safety. The ideal of addressing the needs of early adolescent students was once more forgotten. By the middle 1980s, it was evident that a middle-school life-science curriculum tailored to meet the biological, social, behavioral, and cognitive needs of early adolescents did not exist.

Early Adolescence Today

Students in middle schools (and the remaining junior high schools) today come from different elementary schools. No middle school can be assured that its students will have a common background in science, or even any formal instruction in science. The conventional wisdom among teachers in the middle grades is that, "if you expect children to know some science, it is best you teach it."

In addition, the socializing forces that influence the growth and development of early adolescents today are different from those of past generations (Carnegie Council on Adolescent Development, 1989; Institute of Medicine, 1989). For example, by the time students enter a middle or junior high school they have spent more time viewing television than being in school. Recent changes in family structures and other social factors have resulted in a generation of young people who list lonesomeness as their major problem in life. Stress, depression, suicide, early pregnancy, use of illegal drugs and alcohol, and health problems are on the increase in early adolescent years. The potential for maladjustive behavior in early adolescence should not be ignored in developing a new life-science curriculum.

Science Curricula in Today's Middle Grades

There are two dominant patterns to science curricula in the middle grades (both in middle schools and in junior high schools). One includes a general-science program that extends over a 3-year period. It dates back to 1915, when it was first introduced into junior high schools primarily to motivate students to take more high-school science courses. Topics in the biological, earth, and physical sciences are taught at each grade level in a more or less unified approach. The second pattern includes a series of discrete, year-long, discipline-bound courses in life science, earth science, and physical science. Those courses were developed under the National Science Foundation (NSF) Curriculum Improvement Program of the 1950s, 1960s, and early 1970s. NSF, in conjunction with publishers and schools, is now supporting financially the development of new programs in life, earth, and physical science.

It appears that about half the middle schools have a general-science curriculum and the other half discipline-based courses (Weiss, 1987). In addition, a course in health science is required of middle-school students in most school districts; typically, those who teach it have no training in science.

Teacher and Student Perspectives on Life-Science Courses

Many teachers perceive the purpose of teaching life science as preparing students for the next grade or for high-school biology (Moyer, 1989). They view students as without motivation to learn, deficient in reading ability, and lacking a background in science.

Many students view the life-science course as uninteresting and requiring the memorization of many names. On completion of the course, the typical student reaction has been "never to take another science course unless made to do so."

Textbooks for Life-Science Courses

Approximately 25 life-science textbooks are currently on the market, and they differ little conceptually (Moyer, 1989). Some are written to be continuous with a publisher's elementary-school science series, and others stand alone as a course offering. Their subject matter is essentially a dilution of the traditional high-school biology course. Not surprisingly, they reflect the problems that beset the high-school texts: they are vocabulary-oriented, often containing more than 2,000 biological terms and unfamiliar words.

Conclusions

The middle-school science curriculum is adrift. It is unable to build on a coherent exposure to science in the elementary schools. And it has failed to address the challenge that is implicit in the educational philosophy behind middle schools: to meet the specific educational needs of the early adolescent (Carnegie Council on Adolescent Development, 1989). In a sense, it has simply borrowed from the high-school curriculum, and it is hard to identify any successes.

BIOLOGY IN HIGH SCHOOL

The Importance and Teaching of Fundamental Concepts

At one stage in its development, the study of biology was largely descriptive and based on natural history. Comparative anatomy and the classification of organisms provided vehicles for students to learn a large amount of terminology. For at least a generation, that approach to biology has been inadequate. A mature theory of evolution has stood up well in its broad outlines for more than a century, and recent findings in molecular biology have moved biology to the very forefront of experimental science. Biologists may be on the threshold of tackling successfully two of the most intractable and basic problems presented by living things: how a fertilized egg grows into an adult organism and how a collection of nerve cells learns and remembers.

Biology is a mature discipline underpinned by basic explanatory concepts about how matter is organized in cells and organisms, how genetic information is encoded and transmitted across generations, how parts of organisms are related functionally, how organisms interact with each other and with the environment, and how different kinds of organisms change over time. Students leaving a high-school biology course should have some understanding of those concepts.

The high-school course in biology, normally taken in the tenth grade, is seen by many as the first "serious" science course that students take. Currently, it is also the last formal exposure of many students to science. We have touched on some of the reasons why interest in science is hard to sustain. Previous exposure to science, minimal as it is, has burdened the subject with mystique. Instead of being seen as the way to infer relationships and causes through observation and trial (experiment), which most people engage in to various extents in other parts of their lives without thinking about what they are doing as "science," science is viewed as arcane, difficult, practiced only by the very talented, and unrelated to the real world of the average person. For most students, instead of dispelling those notions, the tenth-grade biology course simply reinforces them. The course also does little to develop scientific reasoning, teach cause and effect, encourage skepticism about correlations and inference, or suggest the value of experimental observation.

Those conclusions are supported by a variety of studies, such as that of O. R. Anderson (1989). Moreover, on the basis of the 1985-1986 science tests for 17-year-olds conducted by the National Assessment of Educational Progress (alluded to in Chapter 1 and discussed further in Chapter 4), Mullis (1989, p. 98) concluded that, "given that most high-school juniors have taken biology, their understanding of the life sciences appears quite limited. . . . On the basis of their lack of knowledge, skills, and understanding and their inability to apply those they do possess, it is likely that our high-school juniors do not grasp the larger concepts that most science educators believe to be the foundation of a strong education in biology, including systems and cycles of change, heredity, diversity, evolution, structure and function, and organization."

In summarizing responses of tenth-graders to items on an attitude scale that is part of the Longitudinal Study of American Youth, Miller (1989) found that only 35% of students agreed that science would be useful to them as adults, 39% were not sure, and 25% said that it would not be useful. Most (59%) of those tenth-graders were enrolled in a biological-science course. Miller emphasized the importance of the biology course in the more general attitudes of students toward science and the role of these attitudes in selecting science courses in the future and in continuing an interest in science out of school.

In an important reflection of the problem of high-school biology, college students (many of whom become teachers) do not fare much better than high-school students. Cronin and Almquist (1988), in a survey of more than 2,100 students on more than 40 campuses, found that 38% of the students polled disagreed with the scientific explanation of human evolution and 45% agreed with a statement that some human races are more evolved than others. Teachers are also a part of the problem. Another study reported that only 12% of Ohio biology teachers surveyed correctly defined the modern theory of evolution and that more than one-third advocated the teaching of creationism in public schools (Epstein, 1987).

Those data—and many others could be cited—indicate that the understanding of science by both students and teachers is deficient and demonstrate the need for more effective teaching and learning at all stages in the educational process.

Textbooks

Textbooks are used by more than 90% of biology teachers (Weiss, 1987, p. 31). In many classrooms, the textbook defines the nature of the course. But most professional biologists who examine high-school texts are appalled at what they find. Most of the texts are far too long and poorly crafted. They contain too much new and unnecessary vocabulary and too little clear exposition of fundamental concepts. They are often boring, and they are also sometimes either misleading or incorrect. In Chapter 4 we explore the writing and marketing of textbooks in greater detail; for the moment, it suffices to say that the textbooks are part of the problem.

Teachers

According to a recent study (Weiss, 1987, p. 40), 76% of teachers are satisfied with the texts now in use. That suggests that a disturbingly large segment of the approximately 37,000 biology teachers are comfortable with a mode of teaching that relies heavily on the use of inadequate textbooks and is obviously failing to serve the needs of most students. The teachers should not bear full responsibility, however. Foremost among the causes (discussed at length in Chapter 5) is the process by which teachers are trained and their expertise sustained through their professional lives. Discussions of teacher training tend to become polarized, with advocates of subject matter pointing accusing fingers at schools of education and charging that the education of teachers is weakened by an overemphasis on pedagogy. As we argue in this report, this is a simplistic view of the problem. The science faculties in colleges and universities have not conveyed the meaning of science to future teachers in a manner that is helpful in precollege classrooms. Nor have they played a sustained and effective role in inservice education. There is ample blame for everyone in the system.

The opportunities for intellectual "retooling" in a rapidly changing science like biology, as well as opportunities for sharing experiences with other professionals (once available to science teachers in summer institutes sponsored by the NSF), have withered. We consider in Chapter 5 the role that such inservice opportunities can and should play.

Conclusions

By any reasonable measure, most high-school students graduate without knowing even the rudiments of basic biological concepts. The students therefore leave school with deep misconceptions about biology that may seriously affect their lives.

3

Curricular Goals for the Near Future

ELEMENTARY SCHOOL

Time for Science

Science—knowledge about the natural world and the processes by which that knowledge is acquired—is a basic subject. To pretend otherwise is to deceive, but we perpetuate the deception in our schools. To present science as a basic subject, we need to make some substantial changes, starting with the time that is allocated to science in elementary school. We believe that during those early years, students should receive science-related instruction. As much time should be spent on science as is spent on the other basics—reading, writing, and mathematics.

Wherever possible, the presentation of science should dissolve the historical boundaries between educational disciplines. Knowledge about the natural world should become integrated with reading, writing, and mathematics. Examples of written science materials that could be incorporated into the normal language-arts lessons include stories about nature and organisms, travel, and how the universe operates. Such reading materials should be written in an interesting style, be filled with visual accompaniments, and be conceptually accurate. Lists of resources of this nature are available, for example, *The Museum of Science and Industry Basic List of Children's Science Books* (Richter and Wenzel, 1986, 1987).

But the suggestion of more time for science involves something much more important than reading and writing about science. The time should be used primarily and most importantly for hands-on exploration by the students. The emphasis should hinge on engagement with, observation of, and direct experience in the natural world. Instruction should not be narrowly focused; it

13

should not suggest at the start of the lesson that something unusual, esoteric, or difficult is about to be studied. The most important goal is to make science attractive and interesting to the students. Many organizations are currently working toward that end, as noted in *Science for Children* (NSRC, 1988), a compilation of curriculum materials, supplementary resources, and sources of information and assistance for elementary-school teachers.

Many districts throughout the country are now developing hands-on science programs for elementary-school children. These districts find it necessary to support their teachers through a comprehensive inservice education program and a science-resource center that supplies modular science kits. At regular intervals, the kits—which include a teachers guide, a set of student activity books, and a set of equipment for the classroom—are circulated from one teacher to another after being checked and resupplied with disposable materials at the district level. Such kits are being designed at the national level in several places, including the Lawrence Hall of Science in Berkeley, California, and the Education Development Center in Newton, Massachusetts. The National Science Resources Center (NSRC) in Washington, D.C., is developing 24 kits of this type: four each for grades 1-6. The first three kits, scheduled for commercial availability in the spring of 1991, deal with plant growth, with electricity, and with microscopes.

Natural History as One Focus

The K-6 years are the appropriate years for developing an "intuitive" (rather than a formal, taxonomic) understanding of biological diversity and the relationships of living organisms. Students should be engaged in observing and caring for a wide range of organisms that can be housed in the classroom, with emphasis on local plants and animals. Many animals can be raised in terrariums or aquariums. Students should assist in feeding and rearing animals to understand their needs, their behavior, and their life histories.

Because plants are especially easy to grow and care for, students at every grade level should be involved with gardening projects, using outside space, window boxes, or potted plants. Both domesticated and native plants should be grown and observed. The ecological and agricultural importance of plants should be a major point of emphasis. The historical importance of agriculture in the development of the human race provides an ideal opportunity to integrate the social and natural sciences.

Students should visit both pristine and disturbed habitats often to observe and study the web of life and how it is influenced by natural and human-related factors. Local resources should be used, such as museums, natural areas, parks, the zoo, and any municipal facility where local talent can be tapped. Local school districts should work together with local resource people to develop appropriate field trips and study sites that can be used routinely by classes. The same sites should be visited periodically throughout the school year to follow the annual weather cycle, thereby stressing continuity and change over time.

Every school should provide transportation to appropriate field sites. A reasonable pattern is two trips per month (on the average) away from the school site and frequent trips around the school grounds or local neighborhood. The self contained classroom model in place in most elementary schools lends itself nicely to scheduling half-day or all-day trips. In contrast, the departmentalized or rotating classroom model in place in most secondary schools makes the scheduling of field trips almost impossible. The enriching experience of the field trip is thus logically the domain of the elementary-school science curriculum.

The Need to Explore

The existing emphasis on learning facts derived primarily from reading is inadequate. It should be replaced by learner-centered lessons that allow students to observe nature directly and practice the skills of inquiry. By inquiry we mean several related processes. Students need to become actively engaged in thinking, asking, and problem-solving. The students' role should be to experience, discuss their experiences with each other, and write about the experiences. The teacher's role is to listen, encourage, ask questions, and lead, but not to act as a font of knowledge, pouring information into empty vessels. Lessons in science should develop skills in careful observation, comparison, measurement, questioning, and communication. As the next step, they should then engage the students in formulating interpretations, conclusions, and explanations. In short, they should reflect more accurately the processes by which science is done and scientific understanding achieved.

Ideally, units of study should be designed as projects, rather than as isolated topics or chapters in a book. During each project, students can produce a tangible product related to the unit of study for display at school and presentation to their families. This approach of "discovery, project, and product" helps each student to develop skills in communication, promotes pride in creativity, engages family support, and develops a positive self-image. The products can include posters, models, photo essays, measurements and presentations of growth or seasonal changes, and the like. Emphasis should be on successful and creative completion of such projects and then on re-examination of results, rather than on worksheets or written examinations, which promote a competitive atmosphere that is detrimental to most young children. Projects also lend themselves well to integrating mathematics and social studies into the units of study.

Achievement Tests

The public's desire to see evidence of school improvement carries a danger. The danger lies not in the understandable wish to see improvement, but in the primitive measures available for assessing what students have learned. This subject is taken up in more detail later in the report (Chapter 4), but a word of caution is in order here.

If education in science in elementary school is to improve, achievement tests must not be allowed to drive the curriculum in wrong directions. Traditionally, such tests have emphasized factual recall and have led teachers to design curricula that "teach to the tests." The tests seriously compromise curricula that are devoted to teaching science as a participatory process and to developing skills in observation and inference.

The California Assessment Program (CAP), an example of a testing program of a different sort, does not monitor specific students or even specific teachers, but attempts to monitor only the overall effectiveness of a schoolwide curriculum. The California *Science Framework* and *Science Framework Addendum* are statewide documents that attempt to define the educational approach and curricular emphases on which CAP testing is based (Science Curriculum Framework and Criteria Committee, 1984). Since the institution of the eighth-grade science test in 1985-1986, test scores have improved each year (results provided by the California State Department of Education). Although the testing program still has far to go to meet the stated goals of an instrument that evaluates conceptual and process-oriented understanding, the attention currently focused on the CAP and a statewide emphasis on meeting the objectives in the State Science Framework are helping to fuel a reform of the science curriculum.

Students should leave elementary school with a strong love for and appreciation of nature and for their own world around them and with the recognition that science is an important way to learn about the world. In elementary school, every student should feel successful in learning science and should look forward to additional instruction. Emphasis should therefore be placed on active participation in science activities, and not on highly competitive grading procedures. These objectives are far more important than either acquisition of the kind of knowledge that is measured by traditional examinations or attempts to identify and reward future scientists. If every student entering the middle-school years already had positive attitudes toward science, the lifelong curiosity about the natural world that would be in place could be exploited in later years.

Science Education of Elementary-School Science Teachers

Teachers of science in elementary school must be far better prepared than are most at present. To disguise their anxieties about science, most elementary-school science teachers have hidden behind textbook-centered lessons that stress vocabulary and memorization of facts. Given the minimal amount of science instruction taken in college by most elementary-school teachers, that attitude is understandable. But the situation must change to achieve quality science instruction in the elementary schools. Because of the breadth required to teach interdisciplinary science well, and because of the very poor science background of most elementary-school teachers, science specialists might be needed to introduce science instruction into most elementary schools.

A science specialist should be trained and certified specifically for teaching science in elementary schools. Such training will require more than the conventional college-level courses in science. The specialist needs training as well in teaching elementary-school science with approaches that engage children in the excitement of the subject. Few programs for preparing teachers currently offer, let alone require, such training for future elementary-school teachers, even those who will teach science.

If elementary schools were to rely primarily on science specialists for science instruction, it could be a major change from the current practice of assigning one teacher to teach everything (except perhaps an occasional lesson on music or art). Such a use of science specialists, however, presents two dangers. First, it could preserve the place of science as a peripheral or supplementary subject, taught only when a specialist is available. Second, it could postpone the necessary preparation of other elementary-school teachers to teach science. Science specialists must therefore be used imaginatively and in ways that do not compromise other goals. One possibility is teaming—assigning a team of two or three teachers with different strengths to teach two or three classrooms of students. Students need not change classrooms; members of the teaching team can move from one classroom to another. Successful models of teaming exist nationwide (see discussion, for example, in the report of the Carnegie Council on Adolescent Development, 1989, pp. 38-40). With appropriate cooperation, science specialists' skills can complement those of other teachers on the team, making them more comfortable with science in the classroom. Furthermore, the skills and knowledge of science specialists can be used in inservice programs to assist their colleagues. We see a role for science specialists in the classroom as a short-term expedient. Preservice education of elementary-school teachers should prepare all teachers to present science with the other basics.

Conclusions

The last 20 years have seen the transformation of the United States into a society that is increasingly dependent on science and technology, but awareness of that reality has not yet permeated our system of education. Science must be treated as a first-priority subject, beginning in the crucially formative years of elementary school. Our general failure to treat it thus is a major reason why secondary-school students perform poorly and harbor negative attitudes about science.

Substantially more time needs to be devoted to science in elementary schools. Biology should focus on natural history, be integrated with other subjects wherever possible, and emphasize observation, interpretation, and hands-on involvement, rather than memorization of facts.

Recommendations

The first three of the following recommendations are starkly worded. Each raises other issues, and each confronts us with numerous obstacles that will have to be overcome before it can be implemented. Most of the rest of this report deals in more detail with how the obstacles can be met and overcome. (In Chapter 8, which addresses the need for national leadership, we propose a means of monitoring and encouraging progress on many fronts.)

• **State departments of education should not only make science a basic subject in elementary schools, but ensure that science instruction is** of adequate quality. This will require much more than token observance of new regulations.

• All programs for preparing elementary-school teachers should institute preservice and inservice activities to assist teachers in presenting science to young children. Likewise, licensing and credentialing boards should require rigorous training of those who will seek to teach elementary-school science. *All* elementary-school teachers should become more familiar and comfortable with science, so that the subject can be truly integrated into the elementary-school curriculum.

• Statewide or district tests should be used only if they are consistent with the goals of a concept-oriented and hands-on elementary-school science curriculum. Achievement tests, when used, should stress conceptual understanding and development of problem-solving skills, rather than acquisition of detailed factual knowledge for its own sake. Inclusion of performance-based exercises in any testing program is also desirable.

• Industry, government agencies, universities, professional societies, and other organizations should assist school personnel and cooperating local resource people in identifying field sites and appropriate field trips to be used by elementary schools. For example, many members of conservation groups and birding clubs have extensive knowledge of local natural history. Local groups could provide summer financial support to help develop programs of study that use those resources in conveying how science is related to the immediate world of the students.

• Attention should be given to the integration of "science stories" into language-arts lessons. Reading and writing about natural phenomena appropriate for the range of readers in elementary school should be an integral part of language-arts and reading instruction. Readings could include stories and other narrative forms and introduce and increase expository science materials over the elementary-school years. Writing should give students opportunities to explain their observations and findings and to examine feelings about the natural environment.

MIDDLE SCHOOL

Human Biology as a Focus

Science courses in middle schools must meet the specific needs and interests of early adolescent students. There are doubtless a number of ways to achieve that goal, but we conclude that the most appropriate formula for the life sciences is one that makes the student the object of study. Rather than perpetuating the life-science course as an anemic version of high-school biology or as a maze of discrete topics distributed throughout textbooks of general science, we would make human biology, broadly defined, the theme. This perspective should raise the student's level of motivation and thereby generate a continuing incentive to learn. In addition, it will provide an appropriate continuation of the science that we propose be taught in elementary schools.

Adolescents at the middle-school level are especially curious about themselves, so links should be forged between the middle-school health-science course and the life-science course. That will necessitate extensive change in the format of both courses. Properly taught, human biology provides a cross-disciplinary perspective on the nature of humankind and what it means to be human. For students, human biology should be learning to know oneself, understanding other human beings, and appreciating their relationships to all other forms of life and to the biosphere. For the teacher, teaching biology means providing a curriculum that not only focuses on understanding oneself, but also increases human potential by developing responsible attitudes about the health of self and others, by reducing maladjustive behaviors (e.g., unhealthy eating and drinking, smoking, and the use of illegal drugs), and by fostering respect for the environment and for the need to sustain a biosphere favorable for the survival of life.

Course Structure

A human-biology course for a middle-school life-science program could be designed as a 2-year sequence that would fill the curricular time slots now occupied by life-science and health-science courses. The subject matter would have both a biological and a cultural and social dimension. The course should include conceptual strands to reinforce ideas of relationships, community, ethics, one's place in the universe, and understanding of self. State requirements regarding teaching about topics related to health and safety should be integrated into the presentation.

Investigative activities should be designed in which students are most often the objects of study—for example, examining human cells, studying genetic and physical diversity in a class population, and studying the local environment.

We must not underestimate the effort that will be necessary to effect the suggested change in the middle-school life-science program, because it cannot be achieved by tinkering. The subject matter for a curriculum in human biology needs to be drawn from several disparate fields of scholarship, and that will require the efforts of others besides biologists. Fortunately, some models are already being developed. One such program is the Carnegie-Stanford human life-science curriculum that is being formulated by researchers from the Stanford University departments of biology, sociology, psychology, and anthropology; the school of medicine (including departments of general practice, psychiatry, and pediatrics); and the Families Study Center (Hurd, 1989a). Representatives of those fields all have teaching assignments in Stanford's Program in Human Biology. Other groups are in the preliminary stages of developing middle-school science curricula that focus on the development of the early adolescent and the inclusion of science-technology-society (STS) themes (NSF, 1988; BSCS, 1989). Another group, the National Science Teachers Association (NSTA), has initiated the Scope, Sequence and Coordination project that addresses science teaching in grades 7-12 (Aldridge, 1989).

Creation of a syllabus is only the start. Courses need to be field-tested, appropriate textbooks must be written, new modes of examination need to be developed, models for preservice and inservice training must be created and tested, and the relationship with health courses and with teachers responsible for that component of the middle-school curriculum must be redefined. (For example, persons not trained in science should not teach science.) The necessary changes are not matters that can be dealt with by casual administrative fiat; they require extensive cooperative ventures by many people who must make the necessary commitment of time and energy, and every school district in the nation will have to be willing to reexamine cherished practices. Many of the potential barriers to implementing curricular change are discussed at greater length in Chapters 4 and 5.

Conclusions

The middle-school life-science course needs drastic revamping. An orientation to human biology holds great promise for both sustaining students' interest in science and addressing a variety of educational goals important to society at large.

Recommendations

- **Several models of refashioned middle-school life-science curricula are being developed. Therefore, the greatest need is not to undertake new initiatives, but to create a process by which programs can be tested, monitored, and evaluated in the science and education communities. As new programs enter the classroom, we will need to know what is working for which socioeconomic, ethnic, and cultural groups; whether a given program has paid sufficient attention to long-term aspects, such as inservice training, and to the development of appropriate testing materials; and what is needed to ensure not only high scientific quality in individual programs, but wide dissemination of the most successful ones. Unfortunately, the argument that good educational programs necessarily push out bad ones rings hollow. Therefore, in Chapter 8 we propose a mechanism for the continuous evaluation and monitoring of science education in the nation's schools.**

HIGH SCHOOL

When this committee began its deliberations, we found ourselves wrestling with the content of the high-school biology course—what it was and what we thought it ought to be. The more we discussed the details of the curriculum, however, the more we saw that we could not convey our vision of the future with yet another syllabus. We fully expect some readers to pick up this document with the hope of finding an outline for the perfect course. They will be disappointed, for our message is vastly more complex than can be conveyed in a syllabus. The high-school biology course, like the other high-school science courses, requires fundamental changes.

Teachers have no shortage of lists and skeletal outlines of topics to be presented in their courses. Those forms of guidance are in fact part of the problem. Putting down yet another bare-bones description of a course will send just the wrong message, for it will invite teachers and publishers alike to look for the topics that they "cover" and, on finding them, to conclude that they must be doing the right thing. Furthermore, academics will complain if their favorite corner of biology is not mentioned. We discuss textbooks in Chapter 4.

We have argued that students who reach the high-school course in biology should already have experienced 9 or 10 years of formal exposure to science. Specifically, we have suggested that the biology to which they were exposed in elementary school should have focused on natural history and that middle school should have helped them to understand themselves as living organisms. By the time they reach high school, teachers should be able to build successfully on that foundation.

We can indicate what the high-school course should be doing by contrasting it with the present version. The explosion of scientific knowledge in the twentieth century confronts us with the need to choose carefully the material to be presented at every level. In making these choices, we must be clear in our own minds about the criteria we are using. All too often, selection is driven by the calendar: If a fact was unearthed last year it must be important. Or if we understand something in great detail, we should teach that detail. The notion of punctuated equilibrium might be but a ripple on the surface of evolutionary theory; because it is current, it has received considerable attention. We know an enormous amount about the molecular details of intermediary metabolism, but to whom are those details important? Certainly not to students with no previous formal exposure to chemistry.

We need a much leaner biology course that is constructed from a small number of general principles that can serve as scaffolding on which students will be able to build further knowledge. Further knowledge can come from reading the newspaper or from course work, but the scaffolding should include an understanding of basic concepts in cell and molecular biology, evolution, energy and metabolism, heredity, development and reproduction, and ecology. Concepts must be mastered through inquiry, not memorization of words. The number of new words introduced must be kept to an absolute minimum.

Examples of What Is Needed

A high-school course with the above aim embodies a substantial departure from the current course. Several examples drawn from different topics will help to convey our intent.

Development and Reproduction

The present course is so burdened with terminology that concepts are lost. For example, the emphasis on naming structures permeates the biology course from molecules to organisms. Consider the process by which the chromosome number is reduced so that each sperm or egg receives a single copy of each chromosome. Many teachers and most textbooks today present the details of meiosis I and II, the specific structures of chromosomes in meiosis I, and the details of meiotic recombination by introducing as many as 20 terms. The names and details of the events during pairing and synapsis are not important in teaching that the process of meiosis halves the number of chromosomes. Students should learn that some chromosomal breakage and rejoining occur and that they increase genetic diversity, but the details can obscure the main function of meiosis—genetic recombination and preparation for fertilization. The teacher could enrich the topic for some students by posing the question of how chromosomes find each other to pair. The students could make their own predictions, inasmuch as the answer is not yet known.

Energy and Metabolism

Many high-school textbooks handle energy superficially. Students must develop an intuitive grasp of the meaning of energy, its different forms, its conservation, and its relation to order and organization of matter in the world of their personal experience before an exploration of energy as the universal requirement for self-sustenance will hold any meaning. The capture and use of energy constitute a common theme that is encountered at the levels of cells, organisms, and ecosystems, but the theme is seldom well developed in the classroom.

Much is known about the details of intermediary metabolism, but ninth- and tenth-grade students need not be burdened with structural formulas of organic molecules and the details of the Krebs cycle. The role of cellular respiration should be developed by focusing on the essential chemistry, involving the stepwise oxidation of organic molecules to form CO_2 and the concomitant reduction reactions in mitochondria—in effect, the charging of a battery that in turn forms adenosine triphosphate (ATP). Expenditure (hydrolysis) of ATP is then coupled to the manifold activities of cells that require energy: making muscles work, building proteins and other molecules, pumping ions out of cells, and so forth.

Finding imaginative analogues in the world of the students' experience and interest is essential. For example, comparison with an engine is instructive, because both boys and girls at this age are developing strong (if superficial) interests in automobiles. Because enzymes enable the reactions to proceed at room temperature, less energy is wasted as heat, and the efficiency of energy conversion to work is high, compared with that in an engine. The formal and reciprocal relationship of cellular respiration with photosynthesis can be developed from the same small number of principles, and green plants lead naturally into the realm of ecology and the Earth's energy and food balance. The latter issues are now usually treated separately, if they are mentioned at all.

Cell and Molecular Biology

Today's students are presented with generalized diagrams of a cell and required to memorize the names of all the subcellular structures, associating each with a function, such as inheritance, secretion, energy production, or digestion. Instead of viewing the cell as a complex of independent factories, it would be more sensible to adopt a functional perspective in which various structures— such as the endoplasmic reticulum, lysosomes, and Golgi apparatus—can be thought of as part of an extra-cytoplasmic region involved in shifting proteins from one compartment to another and out of the cell. In the nucleus, the main concept is that the synthesis of RNA on a DNA template is physically separated from the synthesis of proteins, because the RNA has to pass through the nuclear membrane before it can be translated into proteins. The nucleolus (a prominent structure inside the nucleus with an unfortunate name) is not of special importance functionally for tenth-grade biology students. Other morphological terms, such as centromere and centriole, are merely confusing. Students need to know only that during mitosis the duplicated chromosomes split apart and are moved by the spindle structures to opposite spindle poles. Students need not memorize the names of the various stages of mitosis.

Cells communicate with other cells by a few mechanisms. This important topic is not covered at all in high-school classes, yet it forms a scientific basis for understanding many biological problems, including the action of brain-altering drugs. In many kinds of cell-cell communication, signals must be interpreted by a cell, which then responds in specific ways. These responses occur during signal transmission from nerve cell to nerve cell and from nerve cells to muscles and glands. Other examples are response to light in the retina, response to signaling by hormones, response to artificial drugs (such as opiates), and the response of an egg to a sperm. Although these responses seem very different, they involve a few common mechanisms that start with the binding of molecules outside the cell (the signal) to proteins embedded in the cell's plasma membrane. The cytoplasmic tails of the receptors (the part inside the cell) respond to these external signals by inducing one of a few types of chemical response. In general, chemical responses follow a unified pattern of signal transmission and reception over a time scale from milliseconds in the brain to months in hormonal control of pregnancy. The general pattern is important, but the biochemical details of the chemical responses are varied, often complex, and totally unnecessary to memorize for a student to understand the general significance.

Evolution

The current handling of evolution is egregious. The meaning of the word "theory" has been so corrupted as to spread confusion about the process of science throughout the biology course. As Lerner and Bennetta (1988) have documented, not only do textbooks use "theory" synonymously with myth, legend, or any idea that might pop into the head, but the word is also used as an antonym of "fact." How can a student understand what is meant by "cell theory" or "kinetic theory" when assaulted by such nonsense?

Evolution must be taught as a natural process, as a process that is as fundamental and important in the living world as any basic concept of physics one can name. The evidence that supports evolution—physical measurements of the age of the earth, the fossil record, patterns of similarity in body plans, the records left in the primary structures of nucleic acids and proteins—should all be examined, and students should be led to see how such disparate knowledge knits together to form an elegant and coherent whole.

The existence of evolution should be distinguished from the mechanism by which it occurs; Darwin's contribution was enormous, but its nature should be made clear. Students should understand that natural selection is the principal, but not the only, factor that leads to evolutionary change; they should learn something about the concepts of populations and species; and they should understand the differences between changes that take place in an individual during development and changes that take place in a lineage as a result of evolution.

Evolution is a process and should not be confused with classification, which is a static way of organizing information about organismic diversity. The study of evolution as a process will be most successful if students have acquired some feeling for biological diversity in earlier years through the study of natural history. The study of evolution does not require an extensive knowledge of classification, but knowledge of the evolutionary process provides a framework in which information about systematics will appeal to students. Conversely, dry taxonomic detail by itself is as appealing to learn as the table of organization of a large corporation. Consequently, systematics should appear in the biology course only to the extent necessary to illustrate the process of evolution and satisfy curiosity about the organisms with which most students are familiar. The amount and type of systematic information appropriate might therefore vary, depending on the location of the school and the backgrounds of the students. Our proposals for injecting a great deal of natural history into the earlier years should free much time that is now devoted to systematics in the high-school course.

Ecology

Ecology is often slighted in school; by one estimate, only about 20% of biology teachers find time to treat the subject at all in their courses. The section on ecology often comes at the end of a long text, and classes commonly do not get that far.

Ecology involves connections—between organisms and between organisms and their environments—and students need to develop a feeling for this inter-relatedness as part of their high-school course. We have concern, however, about the effectiveness of the present curriculum with its pervasive emphasis on names and terms. For example, rather than have students memorize lists of every conceivable biotic and abiotic factor on the globe, how much better it would be to engage students in field observations! Rather than have students learn the conventional descriptions of various biomes, we could engage them more productively in analyzing local communities with which they are already familiar. The measurement of microclimatic factors and diversity could serve as the base for studying first-hand the effects of perturbation in similar nearby habitats. Concepts developed from that sort of direct experience will have lasting meaning when they are generalized to the unfamiliar.

Conclusions and Recommendations:
The Scope of Change

The several examples presented here do not constitute a course. They are offered solely to suggest how the teaching of biology needs to change. Skillful teachers will recognize the formula, and many will be able to offer more and better examples. For the most part, however, our schools have little experience in teaching scientific concepts, reasoning, and learning through inquiry; for a teaching force accustomed to lecturing, the demands are imposing:

> Verbal inculcation, however lucid, has very little effect in enhancing reasoning and concept formation. This is not to disparage clear explanation and presentation; it would be foolish to advocate unclear explanation or none at all. Too many teachers, however, labor under the illusion that clear explanation is all that is necessary, and this illusion is a significant source of student failure in development of understanding. Not only is hands-on experience essential, but students must be led to articulate explanations and lines of reasoning in their own words. They must interpret their own hands-on experience, and they must be able to define new terms through appeal to shared experience and simpler words having prior definition [Arons, 1989].

In that brief passage, we have before us the scope of change:

• **In designing a course, we must identify the central concepts and principles that every high-school student should know and pare from the curriculum everything that does not explicate and illuminate these relatively few concepts.**
• **The concepts must be presented in such a manner that they are related to the world that students understand in a language that is familiar.**
• **They must be taught by a process that engages all the students in examining why they believe what they believe. That requires building slowly, with ample time for discussion with peers and with the teacher. In science, it also means observation and experimentation, not as an exercise in following recipes, but to confront the essence of the material.**

We are concerned as much with how science is taught as with the substance of what is taught, and we have considerable doubt about the success of any "reform" that fails to address both parts of the problem. Some will be nervous with this approach, having memories of open classrooms and other educational promises that stumbled. Reform that is perceived as a fad generates disquiet, if not resistance. In such a climate, the slogan "back to basics" might capture the imagination, for its ring of directness and simplicity makes it appealing. But in teaching biology, the conventional promise of "back to basics" is without hope, importance, or meaning. Consider first what is taught—the content. Neither much of the biology of today nor the culture in which it has relevance existed in previous generations, so there is no solid core of basics to which we can be seeking a return. As for how it is taught, it is true that the methods of teaching high-school science have not changed much in the last century; but, as we have explained, there is little cause for pride in that record.

The high-school biology course should instill in students a recognition that science is a process that gives us ways of knowing about the natural world. Students should engage in that process themselves, learning by first-hand experience the skills of measuring and the limits of measurements, becoming acquainted with the practice of reasoning from observation and with the meaning of causation, developing a feeling for scales of size and time that lie beyond direct human sensory experience, and understanding the role of chance in natural phenomena. They should come to see that, although scientific understanding represents our best available analysis and is always subject to revision on the basis of new information, some knowledge is in fact secure and unlikely to change fundamentally, whereas other knowledge is tentative and certain to be refined in the near future.

These goals are more simply stated than accomplished. Inspiring textbooks and skillful teaching are works of art. Like a virtuoso performance on a musical instrument, they require—in various proportions—time, training, practice, encouragement, and inspiration.

In Chapters 4 and 5, we turn to the substantial obstacles that must be overcome before we can achieve the desired goals in a majority of the nation's biology classrooms. We examine there what teachers are taught and how they are taught to teach, how we assess educational success through testing, how textbooks are produced, and what research scientists and university teachers contribute both to the problem and to its solution.

4

Impediments to Implementing Curricular Change: Texts, Tests, and Classroom Practice

Although the committee is concerned about life-sciences education in grades K-12, the focus of this chapter is the high-school level, because this is where the most information has been gathered. Many, but not all, of the recommendations we provide for reform at the high-school level are also applicable at the elementary- and middle-school levels.

TEXTBOOKS

The Present Situation

With estimates that 75% of classroom time and 90% of homework time involve the use of textbooks (Blystone, 1989), it is perhaps surprising that deficiencies in textbooks have not been blamed more for the perceived problems in high-school biology. Nevertheless, there are criticisms aplenty, and the very diversity of their sources has become part of the problem. Criticisms have come from biologists. Several analyses, either anecdotal (Gould, 1987; Paul, 1987) or focused on a specific topic, such as genetics (Cho et al., 1985), have graphically demonstrated how inaccuracies and confusion are perpetuated by textbooks. Textbooks have also been criticized by persons who perceive a challenge to their religious, social, or political views in the treatment of evolution or human behavior. Publishers' attempts to placate religious critics have generated still further negative reaction from biologists.

Are high-school texts truly useful to students and to teachers? Students are discouraged by the overwhelming amount of material and the relentless onslaught of technical vocabulary. The task of learning becomes a treadmill, a daily but endless routine with no useful goals and little sense of accomplishment.

Teachers often have to spend time decoding the texts—a job made more difficult by incomplete explanations and factual inaccuracies. Many students have to rely on their texts, with little guidance from their teachers. Are the present texts capable of sustaining students with little help from instructors?

How might we recognize good textbooks? Biology textbooks are increasingly similar, but too little attention is paid to educationally desirable characteristics. When individual textbooks are reviewed, they are usually compared with each other, rather than with some standard (see, for example, AAAS, 1988); and the general approach, clarity, accuracy, conceptualization, and flow are rarely considered. Criteria for evaluating textbooks have been stated in many forms (see Subcommittee on Instructional Materials and Publications, 1957), but there is general agreement on the following seven needs. Although we are not aware of a thorough recent review based on these criteria, we believe that current texts fail in many respects to meet them.

- *Adequate but not encyclopedic coverage.* Concern about the deliberate omission of such central subjects as evolution and human sexuality is justified, but many current textbooks were written with a compulsion to cover exhaustive lists of terms and topics. The attempt to mention everything—with or without adequate explanations—is a common deficiency of present texts. Its origin is the publishers' correct perception that the current market demands it.

- *Factual accuracy.* This fundamental criterion is often violated. Inaccuracy causes problems for students and teachers alike and raises concern about the qualifications of the authors or the care in preparing the book.

- *Incorporation of current conceptual understanding and new subject matter.* Many publishers add current experimental advances to their texts, often set off in special boxes. But current conceptual understanding generally is not incorporated into the deeper structure of a book and in the specific explanations. Wrong impressions are given. Modern understanding generally simplifies science, rather than complicating it. Good examples are the combination of cytogenetics with formal genetics and, more recently, the use of markers based on DNA structure in genetic linkage analysis. Such subjects are usually taught separately; as a result, the student fails to benefit from the simplifying combination of fields.

- *Logical coherence.* Books written by single authors usually have a logical coherence; multiauthor books often seem like compendia of separate and disconnected segments. If we expect a student to gain increasing mastery of a subject, the subject must unfold with appropriate reference to earlier parts and repetition of older themes. A textbook should be readable and interesting. That is the hardest trait to define and judge, but a common reaction of a knowledgeable person perusing textbooks for the first time is that most fail badly in presenting a coherent point of view and vision of a given subject. The failure is communicated to students, who are understandably bored.

- *Clarity in explanation and effectiveness of illustrations.* Textbook publishers have emphasized illustration, and the amount and elaborateness of illustrated material have increased faster than the number of words in the text. Yet many illustrations are mere decorations—they convey no information, they are often poorly integrated into the text, and they fail to explain.

• *Appropriateness to students' level and interest.* "Appropriateness" has usually been interpreted as not overestimating the students' interests, reading abilities, and capacities for higher thinking, with little thought about positive educational goals, such as understanding of the material. As a result, most texts emphasize rote learning of an alien vocabulary without regard for the likelihood of understanding. As texts have grown in size, the numbers of subjects mentioned have also increased. To make their textbooks suitable for students, publishers should emphasize the teaching of a few important concepts and attend to such pedagogical elements of writing as the proper definition and economical use of terms, the appropriate repetition of important concepts, and the integration of text with laboratory exercises.

• *Representation of biology as an experimental subject.* Textbooks should explicitly convey to students that the information presented is the result of experimentation and that understanding is constantly being refined and is subject to change as new experiments are conducted. Textbooks should also describe the nature of experimentation.

To what extent is the poor achievement of our students in biology a reflection of poor explanations in biology textbooks? Textbooks are only part of a much larger problem, but the lack of sound conceptual bases, the annoying and propagated errors, and the almanac-like organization of material all suggest that the authors' own level of understanding is insufficient to permit effective presentation.

Books need to convey a vision. Furthermore, biology has matured as a discipline, and it is not possible to add new subjects, such as molecular and cell biology, and keep the level of detail on traditional subjects, such as systematics. Similarly, if one wants to teach ecology as a quantitative subject, other subjects must yield. The effort by publishers to mention every subject—aimed at satisfying every conceivable textbook adoption committee—has produced books that do not reflect a satisfactory understanding of any subject. The effect is to introduce numerous barely defined terms, many of which are redundant, imprecise, archaic, and not often used by practicing biologists. To select an arbitrary example, how many scientists now know or care about the specific terms used to identify the morphological stages of meiosis? What is essential is the nature of the processes that are occurring. Similarly, important distinctions can be drawn between endocytosis, pinocytosis, and phagocytosis, but students would be better served by a topological understanding of the cellular traffic of membrane-bound organelles than by memorizing these distinctions.

The most serious deficit in current textbooks is their authors' lack of conceptual understanding of their discipline. Biology now has much of the elegance of classical physics. At the level of the cell, common mechanisms are at work in signaling across cell membranes, in organizing the structure of the cell in cell division or in causing the cell to move, in converting the energy of sunlight into sugars or in the oxidative metabolism of glucose, and in regulating genes in either bacteria or humans. Variation and selection are at work not only in evolution, but in the immune system. Evolution has left its footprints in the structure of DNA and in the strategies of development, and these concepts are important not only in bolstering our understanding of evolution, as introduced by Darwin, but in shedding light on human biology.

The presentation of biology as an experimental subject goes beyond the textbook into the whole curriculum and in particular to laboratory exercises. There is clearly a tension between the demands for textbook comprehensiveness and the limitations of textbook size. The usual casualty is the presentation of biology as an experimental science. In that respect, the books merely amplify the growing pressures of tests and curricula to de-emphasize the process of discovery and to portray biology as the worst kind of literature—all characters and no story.

In summary, current biology textbooks are an important part of the failed biology curriculum. They are often not selective in what they present and lack both a broad conceptual basis and a refined understanding of specific subjects. They emphasize memorization of technical terms. They have many misleading and superfluous illustrations. The books are different, but a tendency toward uniformity and mediocrity can be seen in recent years.

Forces That Shape Textbooks

The open and competitive American marketplace for books might seem to be a guarantee against mediocrity in biology textbooks. Nothing could be further from the truth. In her book *A Conspiracy of Good Intentions*, Harriet Tyson-Bernstein (1988) has lucidly described the interplay of interests at work in the generation of textbooks for public schools. The publishers assert that to survive in the market they must compete effectively in the large "adoption states"—notably California (primary schools only) and Texas—that periodically approve texts for use in their classrooms. Playing on the notion that textbooks, even science textbooks, should conform to local social and religious values, small groups have been able to dictate changes in the content of books before state approval. That phenomenon achieved prominence a few years ago in the case of the treatment of evolution, and recent events in California have demonstrated that it is still alive. Moreover, we are likely to see it in other guises as efforts to teach young children more about human reproduction increase and as animal-rights activists increase their public presence.

Influences from outside the education community, however, are only the tip of the iceberg. Biology books mention large numbers of terms in response to specifications that publishers get from the states. The specifications appear as lists or outlines that are formulated with vague goals or with state or national examinations in mind. Conversely, the examinations might be written with the textbook specifications in mind. Either way, the education community must bear a large measure of responsibility for the characteristics of textbooks.

Publishers are often required to produce a new book in only 6 months or so (Tyson-Bernstein, 1988). In part for this reason, they turn to groups of writers (commonly anonymous), who separately draft small sections of a book. For crafting a coherent textbook, this is a formula for almost sure disaster. Some of the most memorable textbooks (albeit not for high schools) were written by individual authors, for example, E. B. Wilson's *Cell in Development and Inheritance* (first publication in 1896), James Watson's *Molecular Biology of the Gene* (1965), and Linus Pauling's *The Nature of the Chemical Bond* (1939). Each of these authors had a single vision—perhaps idiosyncratic, but economical, unified, and clear. Writing textbooks is one of the most difficult challenges that scholars face, and even some of the most brilliant have failed. It is hard to see how the present manner of writing, often using panels of nonexperts pursuing questionable educational goals, can succeed.

The current method of writing textbooks is illustrated in the preface to the teacher's edition of a widely used high-school biology textbook:

> The modern biology program was developed in conjunction with a thorough program of research and testing. The objective of this research and testing was to survey the wishes and concerns of American teachers of high school biology and to reflect those wishes and concerns in the various components of this program.

The publisher designed the text around "focus groups" that each consisted of

> a moderator and about a dozen teachers from local high schools. The moderator showed the teachers various prototypes for the design of the table of contents, the writing style, and many other aspects of the book. The teachers responded, and representatives of [the publisher] noted the teachers' concerns and then modified the components accordingly.

Note the casual use of teachers, the absence of input from practicing scientists, and the parody of research and testing.

The next step is usually a national market survey. Because study (Weiss, 1987) has shown that 76% of classroom teachers are satisfied with the available books (a serious problem in itself), one is reminded of the commercials for beer in which it is revealed that those who drink a particular brand are found to prefer that brand. There is neither time nor incentive before full-scale production to field-test textbooks under conditions that might expose their weaknesses and lead to revision. (That constraint should be compared with the extensive testing of the biology texts first produced by the Biological Sciences Curriculum Study, BSCS, in the 1960s.)

The number of illustrations, the use of color, and costs are increasing for both college and high-school texts (Blystone, 1989). But to what purpose? Despite the emphasis on illustrations and their obvious technical quality, the pictures often fail to inform. A striking demonstration is the reproduction in the Holt text (1989) of a 1961 *Scientific American* portrayal of the cell, which now appears in color and three dimensions, but with no change in substance. But during the intervening quarter-century, our understanding of cell structure changed dramatically in ways that are useful for understanding cell movement and cell differentiation. Illustrations like this therefore perpetuate incorrect or incomplete concepts.

Illustrations often are not integrated with the text and are not easy to follow. Sometimes they are absurd, as when electron micrographs, which have no natural color, are shown in color. Many illustrations seem to come directly from the scientific literature and are too complicated for high-school students to understand. Does the publisher believe that they give a book an authority that would otherwise be absent? The main function of illustrations appears to be to impress prospective buyers, but in many new texts illustrations are often only decorative distractions.

In summary, most biology textbooks are produced by publishers who are responding to educationally bankrupt market forces. They are written by authors who do not control the content of the books and who are not selected for their knowledge of biology. They are then edited to conform to grade-level readability scores and to accommodate local tastes and religious views. Whatever the educational merits of editing for grade-level readability, even the most casual reading of texts suggests that they are edited by people who know so little of the science that they introduce inaccuracy and confusion. Last, but not least, the current textbooks are not interesting; they fail to convey the fascination and wonder of living systems, thereby convincing many students that the study of biology is an onerous task.

How Can Things Change?

The problem of biology textbooks intersects with a number of other issues discussed in this report. If textbooks are to improve, there needs to be a greater emphasis on the place of concept and process in teaching biology and a clarification of the goals of science education in the sweep of years embraced by K-12. Achieving that emphasis will need consensus among teachers (as described in the section of the report on teacher preparation), changes in how we measure student accomplishment (as described in the section on testing), and changes in the expectations that state boards convey to publishers about texts. If those changes can be accomplished, publishers will find it in their interests to produce better texts.

Improving textbooks is a national problem that requires national leadership. In Chapter 8, we propose the creation of a body that could address the problem of textbooks in ways set forth below.

Recommendations

- A process of review of high-school texts that is open to the scrutiny of scientists should be instituted. Whatever their pedagogical merits, textbooks need to be examined for scientific accuracy, interest, currency, and vision by scientists and outstanding teachers in a forum where the reviews will be widely available to teachers, members of school boards, and others at the grass-roots level. That is, the broader scientific community should be engaged nationally in collaboration with teachers in evaluating textbooks and locally in providing advice in textbook adoption. It is important that teams of reviewers include research scientists, teachers with experience in the school classroom, and individuals familiar with recent research on learning and on reading comprehension.

A fuller examination of present texts for conceptual and factual errors would document further the need for change, enumerate principles that should be stressed in texts, and provide incentives to publishers to alter their mode of production. If conditions can be created in which reviews of books by scientists are truly influential in the processes of adoption, it will become not only possible but necessary for publishers to produce educationally worthy textbooks.

- We need to explore ways in which first-class scientific minds can be engaged in the writing of high-school texts and the control of content can be shifted from publisher to author. We make this statement on the assumption that publishers will welcome serious discussion with scientists. We recognize that good books need not be written by individual authors; in fact, the team approach used by the BSCS, in which research scientists, science educators, and talented secondary-school teachers worked together, has a great deal to recommend it. What is essential is that the right people are involved and that enough time is devoted to the project to allow adequate classroom testing and analysis before books go to press. We could also see a fruitful collaboration develop between a foundation and a publisher for the development of a new text, but if such a project is launched, great effort should be made to see that the mistakes of the past are not repeated.

- A new biology text should be much smaller than most of those now on the market. It should be designed around important biological concepts and principles and should cover fewer topics in greater depth. Moreover, it should be interesting to student and teacher alike. Technical language should be used sparingly and never as a substitute for lucid explanations of biological processes. In the design, writing, and editing of a book, the results of research on reading comprehension should be used fully. Illustrations not only should be accurate, but should be designed to promote understanding. Their conception must be part of the authors' creativity, and they should not be left for editors and art departments to select. Illustrations that only decorate and distract do nothing to train the mind.

• The committee has considered and rejected alternatives to textbooks, such as booklets and videotapes. Although booklets could be topical, could be written by experts, and could be more readily revised than full-length texts, their disadvantages are considerable. The very perishability of loose-leaf or pamphlet formats would impose a cost that most school districts would find unacceptable. Resources that treat selected topics can be very effective in the hands of skilled teachers who are able to provide a conceptual framework to bind the course together, but we are skeptical that most teachers are ready to assume such responsibility for a course of the kind we believe should exist. Textbooks will continue, at least in the near term, to play a central role in most high-school classrooms in the United States.

For videotapes and computer programs the considerations are similar. Although we favor the development of appropriate videotapes and software, we see these materials as playing only supplementary roles in the classroom. Books, however, represent a resource that students should learn to respect and use throughout their lives. In addition, the facilities for using videotapes and computer software are not universally available in American classrooms. Even if they were, there is little likelihood that videotapes and software of suitable quality and quantity could be produced in a short time. In the near term, however, videotapes and computer software could become very effective tools in teacher inservice programs, in addition to their supportive role in the classroom.

This nation has a tendency to try quick technological fixes for national problems, but the solutions to our educational dilemma will not be found in that quarter. Good teachers are the key. We must look to teachers to provide instruction in science to our young people, and in this our teachers need enormous help and support. We must see that they are assisted by the best possible textbooks and not delude ourselves with the hope that merely putting teaching materials in new types of packages or on video screens will prepare our young citizens for the next century.

LABORATORY ACTIVITY

The Importance of Laboratory Activity

Biology offers unique opportunities for students to observe and think about living organisms directly. Nevertheless, the study of life finds students mainly listening to lectures and reading most of each week, week after week (Stake and Easley, 1978; Mullis and Jenkins, 1988; Weiss, 1989). Similar trends at the undergraduate level are adversely affecting the preparation of new biology teachers (Merriam, 1988).

The laboratory serves several crucial functions in the students' intellectual development. First, appropriately designed laboratory activities can challenge students' beliefs about the natural world and lead them to struggle with alternative ideas until they can present scientific concepts accurately in their own words. The effectiveness of that mode of learning was implicit in some earlier approaches to the laboratory, was reinforced by the work of Piaget, and has more recently been broadly supported by findings in cognitive psychology (see, for example, West and Pines, 1985; Linn, 1986; Resnick, 1987). Observation requires hypothesis and theory; without this framework, one does not know what to observe (Frank, 1957). Exposing students' beliefs is important, for naive theories, once exposed, can be made explicit and then tested through further observation and experimentation.

Second, laboratory investigations can enable students to generate knowledge directly from natural phenomena and learn how such knowledge can become reliable knowledge. It is in the laboratory that students can learn the power and characteristics of biology as an experimental science. Laboratory work and field work that are done effectively can enable students to understand scientific ways of knowing and the differences between those ways and other sources of knowledge.

Third, direct hands-on experience produces lasting memory and, if properly reflected on, can lead to deep understanding of organisms and their environments. Part of its benefit is motivation, because it is generally more interesting to study organisms directly than to read about them. Another benefit is engagement. Students can be involved in reading and listening, but participatory involvement with objects and events is more effective.

Fourth, laboratory activity can help students to learn about precision and accuracy in observing, in record-keeping, in measuring, and in inferring. It can help students learn that the need for precision varies with human purpose.

Fifth, laboratories and field studies can involve students in solving problems: defining a problem and stating it as a hypothesis to be tested, determining what and how much evidence is required to make probable or to falsify the hypothesis, controlling variables, learning to use apparatus and techniques to gather valid information, assessing the sufficiency and validity of data gathered, making inferences and interpretations within the limitations of the data, and subjecting the interpretations to the criticism of peers and others (see, for example, the use of discretionary laboratories in Leonard et al., 1981).

Finally, laboratory activities have the potential of introducing students to different technologies—measuring instruments, laboratory apparatus, and electronic equipment—and making them comfortable and skilled in using these tools in the quest for new knowledge. (Budgetary restraints will limit the degree to which this potential can be realized in high schools.)

In science, business, industry, and government, tools and technologies are integral to the workplace. Students must have experience with as many technologies as can be made available in the school setting and, even more important, be provided with opportunities to measure, observe, and infer in solving problems. In summary, laboratory activities can effectively help students to grapple with challenges to their beliefs about natural phenomena and to construct new conceptualizations (Novak, 1988). To attain this overriding outcome, laboratory activities must have personal meaning for each student. As means to this end, laboratory activities should enable students to:

- Formulate problems, devise methods for investigating problems, and solve problems individually or as team members or leaders.
- Deliberate thoughtfully with peers and adults about the outcomes and meanings of investigations and reinvestigate to resolve conceptual contradictions.
- Understand the limitations of small numbers of observations in generating scientific knowledge.
- Distinguish observation from inference, compare personal beliefs with scientific understanding, and comprehend the functions of hypothesis and theory in science and how theories are developed and tested.
- Select appropriate apparatus and use it with skill in conducting investigations.
- Develop familiarity with organisms and an interest in natural phenomena and acquire the knowledge and skills necessary to increase this interest.

Current Failures of Laboratory Instruction

The promise of laboratories has not been realized. The typical laboratory activity is a "cookbook" exercise that students can "do" with little intellectual engagement (Tobin, 1989). Although laboratory work is second to lecture, accounting for about one-fifth of class time in tenth- to twelfth-grade science classrooms (Weiss, 1987—biology was not reported separately), it does not appear to contribute much to student understanding of biology, which is unimpressive (Mullis, 1989).

The use of laboratories in teaching biology is limited and has been declining nationally for years (Stake and Easley, 1978; Mullis and Jenkins, 1988; Weiss, 1989). The decrease has occurred despite the major work done by the BSCS in developing investigative (as opposed to illustrative) laboratory activities, both in the three versions of its biology texts and in its laboratory block and second-course programs (Mayer, 1978). Several factors are contributing to the decline: the decrease in school resources for educational supplies and equipment; teachers' feeling that more time is needed to cover textbooks of increasing length; lack of adequate laboratory experience in teachers' preparation, especially in college biology courses; length of the classroom period;

requirements that teachers move from room to room during the day; the time required for effective laboratory investigations; class sizes that are larger than laboratories and classrooms can accommodate; the lack of space in classrooms to facilitate extended laboratory investigations; inadequacy of funds for buses and substitute teachers for field trips; complaints of other teachers if students are away from their classes for field studies; and the logistics involved in having busy teachers, with too many students and different class preparations, care for and provide materials for classroom use (Stake and Easley, 1978; Hofstein and Lunetta, 1982; Tobin and Gallagher, 1987; Novak, 1988). Finally, laboratory work and field work reduce control of student behavior and can present disciplinary problems, which might be especially difficult for inexperienced teachers or in situations where students do not find meaning in what they are doing.

The goals of laboratory activity are not easily achieved, for without careful design and appropriate discussion of findings, students become confused and discouraged. For example, in a recent study of what constitutes successful laboratory work, Nachmias and Linn (1987) found that students accepted jagged graphs that described results obtained with insensitive instruments as reflecting the nature of cooling and heating. The explanation that the shape was caused by instrumentation failed to influence the students' explanations. Only when the phenomena were explored in depth—with much discussion by students, with student-student and teacher-student questioning, and with explicit instruction designed to help the students to link laboratory information and graphic representation with their knowledge of the natural world—did students develop "a fairly complex understanding of the interrelated factors that influence the smoothness of a heating or cooling curve" (Nachmias and Linn, 1987, p. 503).

Smith and Anderson (1984) came to similar conclusions in their study of experiments on respiration and photosynthesis. After traditional experiments with plants, students retained their naive concepts of food, respiration, and photosynthesis. Ten carefully structured laboratory experiments with extensive and carefully planned discussion and questioning of each experiment were required to achieve conceptual changes in which naive views of food were replaced with biological understanding.

Conclusions

Properly designed laboratory experiences are essential for effective high-school biology courses. The prevalent form of laboratory activities, which merely illustrate what the text has presented, do not produce the desired results and should be replaced with genuine investigations, designed and tested to enable students to achieve the conceptual changes necessary for intellectual development and understanding. Laboratory work and field work are therefore central to a major reconstruction of high-school biology education.

Recommendations

• Laboratory activities will not be able to occupy their appropriate place in the curriculum until time is created to accommodate them. Double periods will have to be scheduled. The difficulties of double periods can be overcome if schools are willing to examine the assumptions that now underpin scheduling of classes. For example, if classes of cohorts are created, science laboratories can be paired with double periods in social studies and physical education. The time required to engage students in science laboratory activities demands that attention be given to this problem by school principals and school districts.

• A major effort should be initiated to identify current exemplary laboratory activities for the high-school biology curriculum. The selection of model exercises should be based on research about how students learn science, should focus on fundamental biological concepts, and should enable students to give their primary attention to developing skills of observation, deduction, and analysis. Due attention should be given to the cost and practicality of laboratory equipment and supplies. Many suitable exercises now exist or could be made more effective with a modest developmental effort. A first approach is to compile and sift. Such organizations as the National Association of Biology Teachers, the National Science Teachers Association, and the National Science Resources Center could be important in this effort.

• A group or groups could also be assembled to develop and assess new model laboratory activities. Such groups should include high-school biology teachers, university and research biologists, and science-education researchers. The inclusion of high-school students (for example, in a summer program) might prove useful or essential to test the activities. The task of such groups should include not only the design and testing of laboratory activities, but also the development of appropriate measures and indicators of the effects of laboratory and field work on student understanding of biology. The participants might be chosen by competitive review of innovative laboratory activities that they have developed or of research relevant to the design of effective laboratory and field activity. Field tests of laboratory activities should be conducted, and revisions made as needed. What the students achieve from the activities and their contribution to the general goals of high-school biology education should be published and made available to all biology teachers, with the necessary intellectual support to place the exercises in the classroom of the typical teacher. Adequate provision should be made for the participation of classroom teachers in the field testing of laboratory and field activities and of the assessment instruments used for evaluating students' achievements. Such teachers should be chosen with the prospect of sharing their experiences in inservice activities with other biology teachers. A program of testing should produce criteria that will enable individual teachers to monitor success in conducting the experimental laboratory activities in their own classrooms and that will assist them in changing their approach when student performance is at variance with expectations.

• A major new program should be developed for providing inservice education on laboratory activities. Such education should enable teachers to be laboratory students and to work through the activities with persons who can interweave the laboratory experience with effective teaching, thus providing a model as to how a particular activity is to be approached most effectively. The assessment instruments should also be used in the inservice programs, modeling how the results are to be used for feedback on instruction and advising about their use and misuse in evaluating the performance of students. Help should be provided on how to integrate the laboratory activities with existing texts and on what teachers should look for in new texts. The program could also develop videotapes to support inservice programs run at the school-district level.

• Animals can be an important part of the laboratory experience, and proper attention must be given to their care. The Institute of Laboratory Animal Resources has provided some guidelines for the appropriate care and use of animals in high-school biology laboratories (see Appendix A).

• A program needs to be instituted for continuing development of effective laboratory instruction and research on such instruction. Biology is a rapidly changing field of study, and effective and significant laboratory activities will need to be created and maintained in a continuing program. The program of development and research should be related to other instructional research, to ensure that laboratory activities are making unique contributions to students' achievement. Priority should be placed on the need to teach students about observation, interpretation, and the processes of science, rather than on the creation of activities based on the most up-to-date technology.

• Programs should be developed to inform teachers and school administrators about the importance of laboratory work and about what constitutes effective and ineffective laboratory work. Providing criteria with which to distinguish effective from ineffective laboratory work will help in defining the role of laboratory activities. Programs should also enable administrators to become advocates for the financial and logistic support required for quality laboratory instruction. Supplies and equipment for laboratory work must become ordinary items in the school budget, as are football helmets and materials for industrial arts. Guidelines for per-pupil costs of effective laboratory instruction would enable schools and school districts to determine the support needed to have effective laboratory instruction in biology. The costs should include the cost of basic equipment, as well as estimated annual outlays for maintenance and consumable materials. Teachers also need some instruction in safe practices and safety regulations, laboratory management, inventory control and maintenance, and the use of computers in course management. These kinds of information should become an integral part of the preservice curriculum. Appropriate documents should be developed for administrators and for use in inservice programs. In Chapter 8, we propose a national process for examining and coordinating various long-term, interrelated needs of science education. Many of the matters discussed here could be addressed in that process.

• The increased threat of litigation is compromising the ability of schools to provide both laboratory activities and field trips. For example, school buses in some districts cannot travel to particular places or cannot carry nonschool personnel (e.g., parents); those restrictions eliminate trips to museums and other sites for science activities. There should be ways to solve that problem, as has been done for athletic teams. School districts, elected officials, and insurance companies must recognize this educational folly and find means to correct it.

• New facilities for laboratory classrooms are needed for both old and new schools, and teachers and other knowledgeable persons should work with school architects to ensure that their recommendations about laboratory design are taken into account.

TESTS AND TESTING

Current Perceptions of Student Performance

High-school biology courses in the United States are generally not producing the kind of understanding that is needed for the world in which today's students will live (see, for example, IEA, 1988; Mullis and Jenkins, 1988; C. W. Anderson, 1989; and Hurd, 1989b, according to whom more than 100 reports on the subject were published in 1983-1988). This conclusion is based on many diverse sources of information. In the preliminary report of its second international science assessment, the International Association for the Evaluation of Educational Achievement (IEA) found that in biology "advanced science students" in the United States ranked last of the students in the 17 countries studied (IEA, 1988). Although the data have received much publicity, they might have serious limitations (Robinson, 1989). That the American students placed substantially below students of the next higher nation raises questions not only about student competence, but about the sample tested and the test itself—questions that might be clarified in IEA's final report.

The National Assessment of Educational Progress (NAEP) has provided the most recent information on how well 9-, 13-, and 17-year-old students in the United States understand science (Mullis and Jenkins, 1988). As described briefly in the introduction, the study contains little that is cause for cheer. Not only was there little difference in performance between 17-year-olds who had taken biology and those who had not, but at the highest level of proficiency— ability to integrate specialized scientific information—students displayed little understanding of cell structures and functions, genetics, and energy transformations. In addition, the data on student attitudes indicate that most students appear to be unenthusiastic about the value and personal relevance of what they have learned, and their interest in science deteriorates as they progress through school.

Standardized Tests

We would like students to take away some understanding from their high-school biology course, but herein lies another problem. Student understanding is usually measured by formal standardized achievement tests. Do these tests measure student understanding of useful knowledge (C. W. Anderson, 1989)? The answer is either "No" or "We really don't know." Furthermore, there is serious concern about the effect of some of the testing practices on the processes of teaching and learning.

A brief digression is necessary here. If most of the test instruments now in use do not assess understanding, how is it possible to conclude that students leave their science courses with little understanding of science? First, the NAEP data (Mullis and Jenkins, 1988) indicate that only 41% of 17-year-olds could analyze scientific procedures and data, and only 7.5% were able to integrate specialized scientific information. The test data of C. W. Anderson (1989) likewise show little student understanding of basic concepts. Other evidence, independent of formal testing, indicates that contemporary biology education is not meeting the changing needs of a society that is now in a global economy characterized by increasing dependence on technology. (See, for example, National Academy of Sciences, 1984; Education Commission of the States, 1985.) That conclusion is based in part on the experiences of members of government, business, and industry as they assess the performance of new personnel (U.S. Department of Labor et al., 1988). New demands, requiring personnel who can handle change in a technological world (U.S. Department of Labor, 1987; Carnevale et al., 1988), dramatize the need for more effective education in science.

Standardized tests use multiple-choice items as the major or sole source of information. Students are asked to recognize the best choice among four or five possible answers to a question. Such tests are limited in determining achievement, in that they measure, by recognition, only a small sample of factual information. Tests that rely on recall do not reflect the conceptual nature of biology, and they reduce and distort testing for concepts by testing for words and definitions instead. Standardized questions that merely require simple recall or name association are of little value to either teachers or students. For example:

The _____ is the powerhouse of the cell.
    a)   cell membrane
    b)   golgi apparatus
    c)   mitochondrion
    d)   nucleus

A "correct" answer to this question in no way conveys an understanding of the important role the mitochondrion plays in cellular respiration.

Standardized tests are generally norm-referenced, which means that they are developed to produce a distribution of scores, preferably a normal distribution. Questions that everyone answers correctly or that everyone misses are eliminated. Questions that are answered correctly by 30-70% of the examinees yield the best "reliability" scores. "Reliability" of a test is the degree to which the same score would presumably be obtained by a person if that person were tested again—it refers to the stability, dependability, and predictability of the test. Reliability is affected by the length of the test; in general, longer tests yield higher reliability. Because norm-referenced tests are designed to sort students on a normal distribution, the mean, median, and mode can shift if students do better or worse, but about half the students will always be "below average." Furthermore, research clearly shows the negative effects of normative grading on learning, motivation, and achievement (Eisenhardt, 1976; Robinson, 1979; Crooks, 1988). There is something pernicious and destructive if pride of accomplishment must be tempered by the belief that one's performance is always "below average."

The contents of standardized tests are developed from the major texts and curricular guides used throughout the United States and are presumably independent of the particular curriculum and instruction of any school or school system. Norm-referenced standardized tests offer efficiency in scoring, yielding numerical scores with high reliability; but they usually sacrifice validity (C. W. Anderson, 1989). "Validity" refers to the degree to which a test measures what it was designed to measure—for our purposes, understanding of science. Standardized tests rarely assess the application of knowledge, problem-solving, or the ability to think through issues that involve biological understanding. The tests, therefore, sort students, but on a scale that deflects judgments about the effectiveness of the curriculum. Students who score well on these tests might discover in college or elsewhere that they have been rewarded without real learning.

An alternate to the norm-referenced test is the criterion-referenced test, which is used by many states to measure competence. Such tests consist of items related to specific objectives and generally incorporate, for a group of items or the whole test, the criterion of acceptable success. Establishing the criterion of acceptability is essentially arbitrary—for example, correctly responding to 70% of the questions. Test items like those administered by the NAEP are scaled to provide a criterion-referenced interpretation of levels on a continuum of proficiency (Mullis and Jenkins, 1988).

Selection of items for either type of test involves judgment, and it would be a mistake to assume that criterion-referenced tests address all the deficiencies of norm-referenced tests. Many items can measure the same concept and vary in difficulty. One cannot assume that test scores indicate accurately the degree to which a concept has been mastered, merely on the basis that the test was criterion-referenced, rather than norm-referenced (Klein and Kosecoff, 1973).

Teacher-Made Tests

Tests can also be composed by individual teachers, and they might or might not include essay or other types of problems in addition to or in place of multiple-choice items. Such tests differ little from those given 30 or 40 years ago, except that fewer essay and more multiple-choice tests are used now. Examination of teacher-made tests indicates that, for the most part, they measure recall of isolated facts in a multiple-choice format (Robinson, 1989; Stiggins et al., undated). In design, tests made by teachers tend to follow the format of standardized tests.

The Educational Impact of Tests

The role of tests has been examined by many others. For example, in an extensive review, Crooks (1988) found that evaluation of students in the classroom appears to be "one of the most important forces influencing education." He found (p. 467) that classroom evaluation guides students' judgment "of what is important to learn, affects their motivation and self-perceptions of competence, structures their approaches to and timing of personal study (e.g., spaced practice), consolidates learning, and affects the development of enduring learning strategies and skills."

A panel of the National Research Council (NRC) convened to review science tests concluded that the multiple use of single tests has led to the misuse of results (Murname and Raizen, 1988). Specifically, the panel recommends that:

- Results from tests constructed for one purpose not be used for a quite different purpose.
- School or classroom average test scores not be applied to individuals and individual test scores not interpreted as ratings or rankings of persons, but only of performance on a test that assesses specific skills.

• Test results or tests of the kind reviewed by the NRC panel not be used as the major force driving curriculum and instruction.

Testing clearly plays important roles in education. Testing can help to motivate interest and generate enthusiasm for a subject, or it can smother interest and deaden curiosity. Testing can help to evaluate the success of programs, or it can drive a curriculum in the wrong direction. We are deeply concerned that the present system confuses the multiple roles of tests, to the detriment of education.

With the recent press for accountability from national, state, and local agencies, testing programs are beginning to drive the curriculum. Teachers feel compelled to limit their efforts in the classroom to what and how school, district, or state tests assess. "Teaching to the test" is increasingly prevalent in high-school biology classrooms, as was vigorously expressed by many teachers in this committee's open forum during the November 1988 meeting of the National Association of Biology Teachers in Chicago.

Dependence on norm- or criterion-referenced multiple-choice tests as they are currently constructed, either for grading students or for analyzing and creating educational policy, tends to misdirect students' habits of study. Rather than mastering concepts, students believe that recognizing terms in a multiple-choice format is the appropriate educational goal. This kind of testing has a major impact on how students go about studying and on the strategies they acquire for learning (Crooks, 1988). In the long run, the impact of current modes of testing on enduring skills and strategies for learning will be inimical to reform. We cannot overemphasize the importance of this relationship.

Resnick and Resnick (1985) make a strong argument for moving from standardized testing to a system of examinations designed to assess particular courses or curricula—examinations created for specific situations. Although in principle such test instruments could be norm- or criterion-referenced, a 1986 project of the NAEP (Blumberg et al., 1986) showed that ensuring that higher-order thinking skills (such as problem-solving) are assessed requires that students be interviewed by knowledgeable adults about responses to problems. Similar results were found in the United Kingdom after more than 7 years of work by the Assessment of Performance Unit (Driver et al., 1964). (For an alternative view, see C. W. Anderson, 1989.)

In the preceding section, we argued that laboratory work and field work are essential parts of an effective education in biology. However, there are few documented instances of the use of practical laboratory tests in biology (Robinson, 1979; Gallagher, 1986). In light of traditional testing practices, it is not surprising that little is known about the effects of laboratory work on achievement in high-school biology. Tamir and Glassman (1970) found differences between student scores on a 2-hour practical examination and teachers' grades; the practical examination seemed to be measuring something different from what was covered in teachers' evaluations. Similarly, Robinson (1969) found a low correlation (r = 0.3) between a laboratory practical examination and a multiple-choice test on similar material. Failure to develop new test instruments to assess properly the outcomes—cognitive and affective, theoretical and applied—of education in biology and especially of the contribution of laboratory work in biology in the current school and university climate could further reduce the commitment of staff and institutions to reform teaching and learning.

Conclusions

Citizens who lack knowledge and understanding are susceptible to those who confuse science and other ways of knowing, to the detriment of public understanding and rational decision-making. Understanding of central concepts and principles of biology will not be gained as long as classrooms and standardized tests assess only recall and recognition. Tests that are consistent with a new commitment to understanding in depth are essential to enable teachers to know what they are accomplishing as they change their teaching methods and emphases. They are also necessary to inform students that different learning strategies are needed to achieve the goals of the biology course.

The fundamental problem faced by evaluators is that they do not have adequate test instruments to determine whether students can use biological knowledge or whether some ways of teaching and learning are more productive than others. For the same reason, we also are not able to determine which major changes in teacher education would be most effective in changing students' achievements. Lack of instruments and methods that permit students to display what they have learned in a year of high-school biology limits the improvement of biology education. In addition, testing is increasingly driving curriculum and instruction in a dull and pedantic fashion, and that makes it imperative to address the issue of testing and evaluation in middle- and high-school biology at all levels—national, state, school district, and classroom.

If major changes are made in the curriculum, in instruction or teacher training or both, existing instruments will be too blunt to show differences in students' achievement in understanding, skills, or attitudes toward science. Continued use of the same kinds of assessment instruments and procedures that are now commonly used could lead to a worsening of the present situation, even in the name of reform. New outcomes and expectations could be nullified by outdated standards.

A whole new array of test instruments and procedures should be developed to enable biology teachers to evaluate and improve their teaching and their students' learning. Means should be designed specifically to address how well schools are doing. Instruments should:

- Serve an educational function for students, helping them to understand their own progress and providing the necessary measurements required by others.
- Inform students of their ability to apply randomly selected biological principles and concepts to real problems—personal, societal, and global.
- Inform communities and the profession about the range of capabilities that students have developed in solving problems, as well as their attitudes toward biology and its relevance to their lives and their future.
- Require the use of scientific apparatus and procedures, such as laboratory practical examinations.
- Serve teachers and administrators in their critical inquiries into the quality of biology education by measuring the range of conceptual capabilities, attitudes, and skills exhibited by students.

Such instruments and procedures should be sensitive enough to display differences in students' performance with changes in teaching and learning. If appropriately developed, such tests might drive the curriculum, but in ways that are in the best interests of students.

Recommendations

Several kinds of tasks need to be carried out by classroom teachers and school districts, states, and the nation. We do not need a national test, but rather tested models and problems that can be used for the assessment of biology education at district, state, or national levels. Testing research and technology have advanced to the point where they can make important contributions to biology education. The American College Testing Program (ACT) and the Educational Testing Service (ETS) seem to be developing new assessment methods, and the trend is encouraging. We have several specific recommendations, as outlined below:

- **A major effort is needed to develop, test, and publish model examinations, as well as sample questions and procedures to assess the desired outcomes of biology education—cognitive and affective changes and theoretical and applied knowledge. The examinations must be especially sensitive to the contribution of laboratory work in biology. They should use any technology that would enable measurement of the most essential outcomes of biology education as described throughout this document. Different tests should be designed for use at different levels—e.g., middle school and high school. They should serve more than traditional functions by involving students in thinking and reasoning (Haney, 1984). Test-makers should consider the work of Blumberg et al. (1986), the Assessment Performance Unit in the United Kingdom (Driver et al., 1964), and other literature on** assessing performance (Stiggins, 1987; Eylon and Linn, 1988). In addition, continuing evaluation of the NABT/NSTA High School Biology Examination (a product of collaboration between NABT and NSTA) with a view to how these goals might be achieved is desirable. The utility and effectiveness of new tests should be examined by analyzing patterns of performance, rather than single numerical scores.

A national group of biology teachers, biologists, teacher educators, and scholars in educational and cognitive psychology and measurement should be funded to collaborate in the development, testing, and implementation of new examinations. Teachers involved in the project should be released from some of their teaching responsibilities to permit direct involvement in this critical effort in instructional improvement.

Once the materials are developed, pilot programs should be created for introducing preservice and inservice biology teachers to their appropriate use. A larger number of biology teachers and college professors who teach methods courses for biology teachers should be involved at this stage. After testing, the programs should be disseminated into many classrooms, with carefully designed research to demonstrate appropriate use and warn of problems that attend misuse of the materials and their adaptations.

• A second initiative is needed to improve tests given regularly by biology teachers during the school year. Unless teachers are helped to construct measuring instruments that are consistent with the best practices and the agreed-on goals of biology education, their classroom tests will lead students to continue to cram for tests that require mere recognition of terms.

This second project should follow the first when there has been enough time for the results of the first to be assessed. During development and field testing, the project should involve classroom teachers, university biologists, and persons knowledgeable about testing procedures. It should include the development of a series of problems designed to assess students' understanding of major biological concepts and ability to apply them. An appropriate outcome would be the creation of a package of widely available training materials that might include a video tape or disk, computer software, and examples of how to analyze patterns of responses to related items.

• The publishers of tests that accompany textbooks and other learning materials need to improve the diversity and quality of their tests. Teachers have come to expect test booklets to accompany textbooks. Because there is no market advantage to testing the quality of the tests, they do not constitute priority investments for publishers, and the schools' ready acceptance of them is a commentary on the low priority attached to testing.

Efforts need to be pressed to enlist the cooperation of university biologists, teachers, science educators, and publishers in improving, through field testing, the assessment materials that accompany textbooks. **Informa-tion on testing conditions, instructional goals, relation to the text materials, and ranges of student performance should be published in standard reports that accompany textbooks.** Unfortunately, given the markets in which publishers swim or sink, we cannot be sanguine about the likelihood that they will meet this challenge until the expectations and practices of teachers and schools are substantially altered.

• **States and school districts should move swiftly to recognize the need for new tests.** Low-quality tests or tests that are not based on appropriate educational goals should not be accepted from publishers. States and school districts should avoid using results of paper-and-pencil tests as the sole criterion of the effectiveness of their biology programs. Tests that are developed locally should go through a careful process to ensure their validity for assessing the outcomes of biology education with respect to essential goals. Several indicators (measures that allow a judgment to be made as to whether a given condition is getting better or worse), as described by Murname and Raizen (1988), should be included with test results when the results are reported. It is essential that indicators be related to student understanding, be operationally defined, and be kept with test results year after year.

The processes of evaluation cannot be addressed in isolation from what is in textbooks, how teachers are taught, the conditions under which teachers work, and public consensus about the goals of science education. Moreover, as science changes, educational goals must be tuned. Biology is a rapidly changing field of study, posing conceptual shifts and generating new social and ethical issues. The issues raised by testing and evaluation cannot be solved without persistent, continuous attention by a broad spectrum of experts.

In addition to the interdependence of testing, textbooks, preservice programs, and inservice programs, there is a second compelling reason to look at testing with a fresh eye: the purposes of norm-referenced testing do not serve the goal of improving teaching and learning. We need a national perspective concerned more with evaluating the effects of curricula, teaching methods, and materials than with ranking the performance of individual students. We need to develop ways to probe the system's components, rather than the relative ranks of the learners. Existing institutions with a role in testing are not designed to pursue that objective by themselves.

In Chapter 8, we propose a formula that is free of constraint from government, business, or any other constituency, but calls on the scientific-research and educational-research communities for valid judgments that will be helpful to biology teachers, parents, administrators, school boards, and state departments of education.

OTHER FACTORS THAT HINDER EFFECTIVE EDUCATION

Effective teaching of biology or any other subject requires a broad and deep knowledge of the material to be taught, instructional skills, a willingness to give of oneself to others, and a commitment that continues after the last child leaves the classroom for the day. In short, teaching is a profession. Among the problems that beset teaching, however, are the nonprofessional burdens that teachers in many schools must bear. Most of those burdens reflect ills in the larger society, but they are often manifested in the classroom, where teachers are not prepared to deal with them. Pernicious problems—such as the sale and use of illegal drugs and the violence that accompanies them, pregnancy of teenagers, and high dropout rates—are becoming endemic in the nation's public schools.

The professionalism of teachers is further vitiated when, in addition to responding to continual crises in the classroom, they are expected to respond to political pressures for accountability and cost-effectiveness. With the increasing public clamor for educational reform, what passes for leadership has all too often fallen solely to politicians, and authority is increasingly centralized in district and state offices of education. That trend has furthered the loss of professionalism among teachers, as decisions about textbooks and objectives have been removed from their control and pressures to teach to examinations have increased. But results of research indicate that the combination of autonomy of schools and teachers with parent participation is the key to high academic achievement (Chubb, 1988), a standard indicator of successful teaching.

A great disparity exists between the goals of teaching and the possibility of reaching those goals in the present teaching environment. The reality of the teaching environment does not bode well for the current wave of educational reform. Within schools, for example, state or local policies often impede effective science teaching. If classrooms have 30 pupils, teachers might have to relate to 150 different children each day. Many classrooms must be shared by teachers, and a teacher must go elsewhere for some periods. Typically, periods are a rigid 45 minutes long, and there is no opportunity during the working day to set up a classroom for laboratory activities. Laboratory facilities, if they exist, are underfunded. Under those circumstances, it is nearly impossible to teach science to students in a manner that is not built around workbooks, lectures, and memorizing.

The absence of time for anything but the minimum makes creative teaching difficult. There is no time to prepare or to reflect and no released time to attend workshops and conferences, to visit other classrooms, or to work with colleagues (even in the same school), to set common curricular goals, to plan together, or to discuss and plan examinations (Taylor, 1989). Teachers are assigned to monitoring hallways between classes or dispensing drinking straws in the cafeteria—activities that can only leave them wondering why they were ever attracted to the profession of teaching. Although decision-makers are generally aware of the extent to which working conditions undermine the professional nature of teaching, they are often unable to change the policies. These policies also convey a lack of understanding of the special role that science plays in a student's education at all levels, as well as the special conditions needed for successful instruction in science.

In addition to the nonprofessional burdens endured by teachers, other aspects of the teaching environment impede effective education. The classroom has become a microcosm of the myriad social and economic problems of society. Teaching toward the goal of academic excellence is but one of the many tasks that teachers are expected to perform. Schools are expected to serve other purposes, such as socializing young people into their culture and preparing students for specific occupations. But life experiences that students bring into the classroom affect how receptive they are to learning. Teachers are expected to help students overcome their nonschool problems. An endemic culture of poverty in the inner cities of large urban centers perpetuates harmful social activities. Moreover, peer pressure often discourages students from even trying to succeed academically (Chubb, 1988). A recent report of the Institute of Medicine brought to light the prevalence of mental illness among our nation's youth (Institute of Medicine, 1989). Inadequate pay provides little incentive for teachers to stay in the teaching profession in the face of these and other obstacles, such as overwork and the violence directed at teachers and students in large urban areas. Teachers must also overcome the negative image of science held by students and often by their parents.

Overcoming the barriers to effective education is often left to the teachers, who are helpless to effect any major change, given the dynamics of a system over which they have little control. Instead, they are forced to work within a hierarchical system imposed by both school systems and teachers' unions. One example involving salary will illustrate the point. Except for states with a single, statewide salary schedule, each school district sets remuneration with the usual provision that teachers will not be hired into the district at a salary that recognizes more than 6 years of experience. Therefore, senior teachers who leave a school district must take salary cuts to move. In effect, unlike members of most university and business professions, public-school teachers who have stayed in a school district (or state) for more than several years become permanently indentured servants. The members of no other profession are so hobbled.

If the goal of effective education is to be realized, teachers must be given more control over the system in which they work.

Recommendations

The following recommendations will expedite reform by increasing the professionalism of teaching.

• **Seriously addressing the need to teach science more effectively will require changing some current administrative practices.** More flexibility is required in the scheduling of classroom and preparation time, in the pursuit of related professional activities by teachers, and in teachers' sharing of responsibilities.

This recommendation carries some inevitable fiscal implications, and a word on that score is appropriate. Goals of "reform" are of two kinds. The first merely specifies minimal accomplishments that can be effected by policy changes or new rules. The second aims higher by attempting to alter the result of the educational process in some fundamental way, e.g., by maximizing the intellectual accomplishments of individual pupils. Requiring more courses for graduation, lengthening the schoolday, and inserting yet another normalized test are examples of the first. Although they might seem to speak to the problem, by themselves they are merely political palliatives that leave the impression that something important is being done. In contrast, changing the outcome of schooling in a basic way seriously disturbs the system and presents a challenge to virtually every interest group on the scene—teachers, administrators, and taxpayers (Airasian, 1989). True educational reform will be expensive and rock many boats, and those who must pay for change should be clear about the goals they wish to achieve.

• **Obstacles to effective teaching must be lifted.** Inasmuch as textbooks and testing play major roles in determining how biology is taught, teachers must be encouraged to experiment with new techniques of pedagogy and assessment. School policies, rather than perpetuating isolation, should be tailored to support and encourage teachers in working together in developing ideas. School administrators can endorse activities for teachers to share ideas about new curricular approaches through policies allowing released time and "free" time during the school day. Teachers must also be encouraged to exchange information about what works and what does not work in the classroom—e.g., pedagogical techniques and laboratory approaches. School policies should encourage teachers to become involved in new curricular projects and should assure them of long-term commitment and support for successful innovative efforts.

• **The nonteaching tasks assigned to teachers should decrease.** Teachers should be expected to devote nonteaching time to activities that will enhance their ability to teach, such as laboratory preparation and tutoring students. Valuable time should not be spent in monitoring hallways or supervising lunchrooms.

- School boards must be convinced that hiring experienced teachers, who are paid more than less-experienced teachers, is sometimes best for the children in the district. Market forces would then play a greater role in public-school education. Districts that provided the best working conditions for effective teaching would be able to attract the best teachers. Other districts might have to improve working conditions and salaries, if they are to retain or attract highly qualified teachers. Through such a change in the conditions of teacher employment, both teachers and school boards would have freer hands in creating faculties and working conditions that lead to schools that communities would wish to support with enthusiasm.

5

Impediments to Implementing Curricular Change: Training and Support of Teachers

PRESERVICE EDUCATION: TEACHING THE TEACHERS

The present process of learning to teach is long and continuous. It begins well before a prospective teacher enters a formal program, during 12-14 years of "apprenticeship of observations"; by watching teachers go about their work, students garner many ideas about teaching and learning (NCRTE, 1988). Formally, it starts with admission into a program for educating teachers (*preservice* education), usually in the sophomore or junior year of undergraduate education. It continues with *induction* into the profession (practice teaching or interning) during the senior year or in a postbaccalaureate year and *inservice* education throughout one's teaching career. When the student completes the preservice program, state policies determine what subjects one may teach at what level and for how long.

Preservice Education of High-School Biology Teachers

The undergraduate biology and related science education of prospective American teachers varies greatly, depending on the type of institution they attend. Traditional programs for educating biology teachers at universities that have schools or colleges of education include a biology major and pedagogical and professional courses that can be completed in 4 years. A successful graduate receives a bachelor's degree and qualifies to apply for a teaching license. More recently, some research universities have moved to programs that couple a standard biology major with a fifth (and sometimes a sixth) year of pedagogical, professional, and practical experience. Most programs requiring 6 years result in both bachelor's and master's degrees, as well as teaching licenses; 5-year programs lead to a bachelor's degree and a teaching license. New York and California have required 5-year programs for more than 50 years.

Generally, undergraduate courses taken by preservice biology teachers are the same as those taken by students preparing for professional or graduate schools. Two professional associations, the National Association of Biology Teachers (NABT) and the National Science Teachers Association (NSTA), have recommended numbers and types of courses for programs instructing future biology teachers (Appendixes B and C).

Our committee is concerned about both the scope and the format of science courses currently available to prospective teachers. In most colleges, large, impersonal lecture courses and structured laboratory activities in science departments make up the format, and prospective teachers have few opportunities to participate in long-range laboratory inquiries, to lead fruitful discussions, or to ask and respond to penetrating questions. The infrequent use of creative inquiry or of strategies for cooperative learning in high-school biology classrooms is probably related to their absence in most college programs. A recent assessment of mathematics education (NRC, 1989a) came to the same conclusion in suggesting that a major problem in elementary-school and secondary-school mathematics instruction is that most teachers have studied only in an authoritarian framework.

The component of preservice course work that deals with teaching methods involves general instruction in the sociology of schools and the psychology of learning and usually a one-semester course in methods of teaching science. The fundamental difficulty with the courses in pedagogy is that they are unrelated to the specifics of teaching the concepts of biology. Skillful and experienced teachers have discovered, through practice, effective techniques for teaching specific scientific concepts, but little of this useful information is incorporated into the preservice education of teachers. Recently, the curriculum has been augmented with courses on, for example, early field (classroom) experiences, multicultural education, and science, technology, and society (STS). But few institutions include courses on research about learning, on building and practicing techniques of communicating science to students at different ages (content pedagogy), or on inducting new teachers into the schools. The courses dealing with both content and pedagogy need to be changed; however, the most pressing need is to integrate these two components of preservice education.

Preservice Education of Elementary-School, Middle-School, and Junior-High-School Teachers

Programs to educate prospective elementary-school and middle- or junior-high-school teachers to teach science are even more diverse and inadequate. Historically, preservice education for elementary-school teachers (grades K-5 or K-6) has prepared generalists who have extensive training in pedagogy. Most prospective elementary-school teachers major in elementary education in colleges of education, and their undergraduate programs generally consist of one-third general-education courses (such as English, speech, and psychology), one-third content courses (such as language arts, children's literature, biology, physical sciences, and mathematics), and one-third pedagogical courses (exceptional children, methods of teaching in each subject, classroom management, and so forth). Depending on the institution, elementary-education majors take regular courses in a subject (e.g., introduction to biology) or courses especially designed for them (for example, biology for elementary-education majors).

One feature consistently characterizes those programs: prospective elementary-school teachers study very little science. As a group, they are therefore poorly prepared to teach science, and most of them devote little time to science instruction in their classrooms. Weiss (1987) reports that in 1985-1986 the average time per day spent in teaching science in elementary school is 18 minutes for grades K-3 and 29 minutes for grades 4-6.

Preparation of prospective teachers in middle or junior high schools is not based on a coherent philosophy. In some states, teachers for middle and high schools come through the same kind of preservice programs. In states that have special requirements for licensing for grades 6-9, however, institutions have developed preservice programs for middle- or junior-high-school teachers. Usually, such programs are in colleges of education; that is, students major in education, but have a minor in science. One common weakness is that many students take survey courses in several sciences and do not gain in-depth preparation in any one science.

The Process of Induction in the Education of Precollege Teachers

The final preparation for teaching in all grades is student teaching, which is usually a full-time experience for less than a semester. Recently, longer (and partially paid) internships have become more popular in graduate schools of education, which collaborate with what are called professional-development schools. Such schools might be jointly operated by school districts and universities, with the aim of providing a more structured and supportive introduction to teaching than is usually available.

Internships can replace student teaching (usually in 5-year programs) or be used in a special first-year position in which beginning biology teachers are under the supervision of an experienced mentor. In other cases, internships are part of a program in which a university warranties the quality of its education graduates by agreeing to assist first-year teachers (often called interns) in overcoming deficiencies. Such warranty programs are available in a variety of institutions, such as research universities, state universities and teacher education institutions, several small colleges, and some institutions predominantly for blacks.

Currents of Reform and Their Possible Impact

Several reports have recently refocused national attention on the education of teachers. The Holmes Group (1986), a consortium of more than 120 universities with a commitment to research, has developed an agenda for training teachers that is radically different from most existing programs. It proposes to abolish the undergraduate education major; to develop a differentiated teaching force, distinguished by specific disciplinary teaching licenses; and to establish programs for professional development and practical training in which prospective teachers would enroll for a fifth year after receiving their bachelor's degree in a 4-year liberal-arts curriculum.

The Carnegie Task Force on Teaching as a Profession (1986)—a group of industry, government, and education leaders—has specified similar changes for programs that educate teachers. In addition, it has supported the establishment of a National Board for Professional Teaching Standards that would restructure the teaching force into four levels. The task force further recommends the development of clinical schools for the practical component of teachers' training. The Holmes and Carnegie groups differ on some issues, such as the urgency of recruiting minority-group members into teaching and the importance of assessing students' learning and teachers' accountability. Together, however, they provide both a focus and an agenda for reform.

The education of teachers in the United States appears to be entering another period of substantial change. Although it holds promise of improving the programs in which teachers are trained, effective reform will have to occur *within* the broad spectrum of institutions that educate teachers. Perhaps because of their exclusion or lack of involvement, an association of institutions traditionally devoted to the education of teachers has questioned the recent suggestions for reform. The Teacher Education Council of State Colleges and Universities (TECSCU), a group of institutions that has a long tradition of preparing teachers for elementary and secondary schools, supports retaining the education major and the standard 4-year program, requiring the same standards for certifying all teachers, eliminating the establishment of national standards of competence, and strengthening the role of the school principal.

Much of the current reform movement focuses on the comparative efficacy of degrees in education and degrees in the arts and sciences, but little empirical research informs the debate. What teachers actually learn from science or education courses and how that learning is related to their effectiveness in the classroom are not well understood (Guyton and Farokhi, 1987; NCRTE, 1988). One of the current problems is the inability to sort out the effects of informal observations, practical experiences, and formal courses on what a teacher learns and how a teacher teaches. Several current studies might help to clarify the matter (Hummel and Strom, 1987; NCRTE, 1988).

Most programs, in both science departments and schools of education, lack a continuous series of experiences that allow prospective teachers to develop skills in using analogies and examples to illustrate and clarify the science to be taught. The Holmes Group initiatives do not explicitly address the development of pedagogical knowledge that is specific to a subject; rather, they assume that knowledge of content (e.g., gained through a biology major) is adequate. An add-on fifth year might separate pedagogical courses further from content courses and will not provide continuous structured opportunities to integrate pedagogy and content. The issue is critical, because the pedagogical techniques for teaching specific science content as a process of inquiry are largely missing from all present college and university curricula.

In the preparation of elementary-school teachers, two routes have been taken by institutions adhering to the Holmes Group recommendation: some have developed special interdisciplinary majors (e.g., science-mathematics, language arts, life studies), and others have steered prospective elementary-school teachers into traditional majors. With the elimination of the elementary-education major, one concern is that few prospective elementary-school teachers will choose the science-mathematics interdisciplinary major or a traditional major in one of the natural sciences. The science requirements in many humanities, arts, and social-science majors are even lower than those in current elementary education majors, so the next generation of elementary-school teachers might have less experience with science than do the current teachers.

As part of a Carnegie study of adolescent development, a task force has advocated special preservice programs for prospective middle-school teachers that provide a firm foundation in science (including depth in one field), courses on early adolescence, and pedagogy integrated with practice (Carnegie Council on Adolescent Development, 1989). In one suggested approach, prospective middle-school teachers would enroll in traditional academic majors, which would be augmented by opportunities to observe and work with early adolescents as early as the freshman year. The task force recommends an undergraduate education with a concentration on two academic subjects followed by a period of internship or apprenticeship in a middle school (Carnegie Council on Adolescent Development, 1989).

The National Board for Professional Teaching Standards, initiated by the Carnegie Task Force on Teaching as a Profession (1986), has made recommendations for preservice education of teachers and stresses the need to integrate subject and pedagogy throughout the preservice program.

Conclusions

The preparation of future teachers is in need of drastic reform. Existing standards for both content and pedagogy are inadequate to meet current societal expectations. The problem will be exacerbated in the next decade, unless much stronger teacher preparation is initiated. Effective biology teaching requires being able to do, as well as to know, and new programs must ensure that new teachers not only understand biology, but have the skills to relate scientific concepts to children of different ages.

Recommendations

Our recommendations for preservice education and the induction of teachers focus on the nature of courses and programs in the undergraduate major, the need for appropriate research on teaching, and the type of institution for educating future biology teachers.

- **A high-school curriculum that treats science as a *process* for knowing about the world can be successful only if the teachers have a deep understanding of that process themselves. We therefore feel that every teacher who has responsibility for a high-school science class should have** had the experience of engaging in original research earlier under the direction of a research scientist. Ideally, that should happen as part of preservice education, even if for only a semester or a summer. For active teachers who have missed the opportunity, inservice mechanisms must be devised, as recommended below.

• Prospective teachers of high-school biology should be prepared in cell and developmental biology, ecology, evolutionary biology, genetics, and molecular biology and biochemistry. Those fields should guide their selection of courses, which should be underpinned by a basic exposure to mathematics and the physical sciences. We encourage experiences that explore new ways to break down traditional barriers between the natural sciences and between the natural and social sciences. Wherever possible, the curriculum should include at least one course in which science is related to issues of public concern.

• The most important change in the undergraduate curriculum will require the participation of university science and education faculty in creating environments for learning that are less authoritarian and that engage future teachers in discussion of concepts, the relations between scientific disciplines, and cooperative analysis of information. New, more effective processes should be developed for integrating pedagogical and scientific subject matter. In schools that train many teachers, special sections could be created in which the students have an opportunity to discuss how their experiences at the college level could be best modified to present important concepts and principles to younger students.

• Faculties of schools of education and science departments should collaborate to develop science-methods courses. The goal of such courses would be to combine appropriate teaching pedagogy with scientific methods; they would be taught by biologists or biology-education specialists.

• Undergraduate programs are needed that will better prepare teachers to deal with science in elementary and middle schools. Such programs could have an integrated science or science-mathematics major. The pedagogical character of the programs will differ from that appropriate for high-school teachers, but there are few if any usable models.

• Other issues that need attention include instruction in the strengths and weaknesses of different procedures for testing at different age levels and evaluation and selection of appropriate curricular materials, especially textbooks.

• There is a dearth of research on what makes teacher-education programs effective. Among the questions that need to be asked and analyzed are the following:

—What is the relative efficacy of 4- and 5-year programs?

—What scientific skills, strategies, and knowledge are most needed by biology teachers?

—What facilitates the acquisition and use of those skills and knowledge by novice teachers?

—What type of induction period maximizes teachers' effectiveness and students' learning?

- Some institutions now training biology teachers should not train them. For example, some fundamentalist colleges and universities do not teach evolution. A national group should consider the material that must be offered to provide adequate preservice education in the sciences.
- Current movements for reform of the teaching profession argue that adequate preservice education requires more than 4 years and a bachelor's degree. We support that view, but feel that several patterns in 5-year programs might be valid. For example, a student who receives a bachelor of arts or science in biology as part of a liberal-arts curriculum could spend a fifth year largely under the wing of a senior mentor, obtaining experience in the classroom, rather than in a full program of more courses. Regardless of the details of the preservice experience, however, teachers' education should include both attention to "content-pedagogy" and carefully designed inservice programs. Particularly during the first several years of teaching, the focus of inservice programs should be on techniques for teaching science. In later years, they should provide mechanisms for updating teachers' knowledge of science.
- The plans of the reform movement to lengthen preservice education from 4 to 5 years are likely to make it more difficult to attract talented people from impoverished backgrounds and have a disproportionate impact on minority groups. As part of recruitment, institutions must find ways to allow prospective teachers to fulfill undergraduate biology majors and teacher licensing requirements without additional expense.

LICENSING AND CERTIFICATION OF TEACHERS

Until recently, the terms "license" and "certification" were used interchangeably to indicate the legal approval by a state to teach, but today there is a distinction between the two terms and the two processes. States grant licenses; professional groups confer certification. Governor Thomas Kean of New Jersey articulated the difference (National Governors' Association for Policy Research, 1988, p. ii):

> Board certification will be different from state licensing. State licensure will continue as a prerequisite for teaching, while professional board certification will offer an option to teachers who would like to be recognized for what they know and can do as first class practitioners. Both will exist in parallel as they do in most other professions.

Licensing

When students complete approved course work and practical experience, they receive short-term (3- to 5-year) state licenses. Licenses are for specific subjects and teaching levels. Traditionally, states approve college programs and graduates receive licenses automatically. State requirements include type and number of hours in specific disciplinary and professional courses, as well as general competence. In the case of biology, most practicing teachers are licensed. A survey of 23 states indicated that only 9% of newly hired teachers and 6% of practicing biology teachers did not hold licenses to teach biology as a major or minor field (Champagne and Baden, 1988). However, there are shortages of teachers in the physical sciences, and chemistry and especially physics are often taught by persons without appropriate licenses. That situation is particularly common in rural schools, where only one or two classes of chemistry or physics are offered each year. In some states, superintendents can ask for emergency licenses that allow teachers to teach out of their own fields.

Subject specialty is often ignored in the assignment of teachers to science classes in middle and junior high schools. Many teachers with specialties in agriculture, home economics, or general science teach whatever middle- or junior-high-school science is offered. A license to teach general science does not mean that its holder has adequate training in biology, chemistry, physics, and earth science; many persons with this category of license have only a general educational background in science. Teaching licenses granted in one state usually are recognized in another. Most colleges and universities seek approval of their programs for educating teachers by the National Council for Accreditation of Teacher Education (NCATE). NCATE's imprimatur is important, because graduation from an NCATE-approved program assures prospective teachers that their licenses will be honored in most states. NCATE standards have emphasized pedagogy, often to the neglect of content. Moreover, many preservice programs in research universities have not met NCATE standards for the number and variety of field experiences or number and kinds of education courses.

Recently, two professional associations of teachers, the National Association of Biology Teachers (NABT, 1985) and the National Science Teachers Association (NSTA, 1984), have recommended appropriate courses and teaching competences that have ramifications for both licensing and certification (Appendixes B and C). NABT recommends criteria for minimal content and competence in an undergraduate program leading to a biology-teaching license; NSTA has developed mechanisms to review preparatory programs, to evaluate inservice education, and to bestow approval on teachers. The recent adoption of NSTA standards by NCATE makes those standards potentially important. Weiss's (1987) survey results indicate that 80% of responding high-school biology teachers meet the general 32-hour biology requirements suggested by NSTA, but only 29% of all biology teachers satisfy all the NSTA requirements; the rest generally lack one or more specific courses.

Until recently, graduation from an approved program has been sufficient for a teaching license. During the last 5 years, however, national examinations have been added. In general, they attempt to assess both basic skills and professional knowledge. Today, 42 states use tests to screen beginning teachers, and the remaining eight plan to implement them soon. Of the 42, 22 require the National Teacher Examination, offered by the Educational Testing Service (Champagne and Baden, 1988).

The increase in the use of tests to measure the competence of teachers is part of the current reform movement. Tests (or grade-point average—GPA—in courses) may be used on entrance into preservice programs or later, as part of the licensing process. Reliance on those criteria, however, raises a number of issues that have not been well explored. Guyton and Farokhi (1987) studied whether basic skills and academic performance were related to subject knowledge and teaching performance in the classroom. According to the authors, the study failed to support three current trends that have led to the adoption of examinations for teachers: testing of basic skills before entry into teacher education does *not* screen out persons who will become less-able teachers; the GPA is *not* a predictor of a teacher's performance, so raising the GPA as a requirement for entrance into teacher-education programs is of questionable value; and it cannot be assumed that one can equate subject knowledge with ability to teach. Analysis points up the need to relate pedagogy more effectively to subject matter in the training of teachers and the need to examine critically new procedures that purport to assess teachers with examinations. However, the GPA is not irrelevant as a measure of academic performance and we should try to attract talented individuals to the profession of teaching.

The State of Georgia has sponsored studies to assess teachers. Using lesson plans submitted by the teachers and observation of classrooms, the state has invested over $2 million in developing a system for evaluating the performance of new teachers (Bethel, 1984). Other states are following Georgia's example. Teaching licenses in many states are tied to passing the assessment process, but critical evaluation of the process itself is still lacking.

Perhaps because teacher licensing is cumbersome and burdened with specific regulations, reformers have sought to implement and test alternative routes to licenses. In 1987, 24 states had alternative licensing programs designed, at least in part, to increase the numbers of science and mathematics teachers by decreasing the number of disincentives to entering the teaching profession (Blank, 1988; see Appendix D). Those programs allow college graduates to become teachers without matriculating in a formal preservice program.

One of the most publicized new routes has been the Provisional Teacher Program approved by the New Jersey State Board of Education, under which anyone who has a baccalaureate degree, passes an examination in a particular subject, and completes a 1-year intern program may be licensed to teach (Cooperman and Klagholz, 1985). Furthermore, the program permits a school district to select a well-qualified, nonlicensed provisional teacher over a less-qualified, licensed teacher. Since it began in 1985, more than 1,500 new provisional teachers have entered New Jersey classrooms through the Provisional Teacher Program. Nearly 30% of the state's new public-school teachers in 1988 and 24% in 1989 were hired under the program (New Jersey State Department of Education, 1989). State officials claim that the caliber of new recruits is on the whole higher than that of the state's licensed teachers (*New York Times*, September 13, 1989). New Jersey's experiment needs to be followed closely. One potential problem is that teachers unions view alternative paths to licensing with suspicion, seeing them as mechanisms to keep teachers' salaries low in periods or areas of high demand and low supply.

Certification

Professional certification of teachers is one of the cornerstones of the current reform movement. The first professional organization to grant certification was NSTA. Both the Holmes Group (1986) and the projects initiated by the Carnegie Corporation of New York (1986) advocate professional certification of well-prepared teachers. The Holmes Group envisions differential certification of graduates of 5-year programs of teacher education. The Teacher Assessment Project, funded by Carnegie at Stanford University, has been developing standards that could be used to confer professional certification. Various interrelated programs in the Connecticut Continuum project also have potential as a means to this end. The National Governors' Association and the research and development projects funded by Carnegie have supported the recent establishment of a National Board for Professional Teaching Standards (NBPTS) that eventually plans to certify teachers who have fulfilled special preservice and induction requirements. Practical skills, pedagogical skills, and pedagogical knowledge of the subject would be assessed, in addition to subject knowledge. One basis of the project is to engage teachers in the setting and meeting of standards for membership in the teaching force. The NBPTS therefore has drawn two-thirds of its members from the teaching profession and one-third from the public and from universities. The NBPTS expects to begin certifying teachers in 1993 and hopes eventually to certify every qualified teacher in the nation. Both the suggested requirements and the process of certification make the implications of this program profound, but acceptance and support by teachers are as yet unknown.

If adopted by states and teachers unions, certification will force the following changes in the education of biology teachers:

- Induction into teaching that starts during preservice and extends over several years.
- Opportunities for developing skills in teaching science combined with opportunities for reflection on teaching effectiveness.
- Teaching as a shared practice.
- National examinations in subject matter, which will influence courses in preservice programs.
- Inservice programs in which practicing teachers and principals will serve as mentors.

• Cooperation of colleges and universities, schools, and unions in helping candidates to meet requirements for certification.

Conclusions

In efforts to improve the performance of our nation's schools, attempts are being made to strengthen the licensing process and to create an alternative in the form of professional certification. Changes in licensing requirements have so far focused on examinations of debatable relevance and on alternative licensing schemes that hold considerable promise, but are also subject to administrative misuse. Plans for certification have the potential for creating generally accepted national standards.

Recommendations

• **State licensing regulations should be altered so as to facilitate, not impede, improvement of preservice education.**

• **The various alternative ways to obtain teaching licenses (late entry, long internship, etc.) should be critically evaluated. Existing models, such as New Jersey's Provisional Teacher Program, need to be compared, particularly with respect to the teaching of science.**

• **Questionable routes to licensing, such as emergency certification and seniority rules that cut across disciplines, should be eliminated. The practice of assigning unqualified persons to teach science simply because they have seniority in the school system is without educational justification and must cease.**

• **We support a thorough review of the National Teacher Examination to ensure that it is related to teaching performance, not simply to basic content knowledge and exposure to required courses.**

• **Major changes in certification along the lines of Stanford's Teacher Assessment Project or the Connecticut Continuum project are occurring. An independent national committee—composed of biologists, teachers, biology educators, and state school personnel—should evaluate those plans and others with particular emphasis on their adequacy for assessing competence to teach science as a process of inquiry and discovery.**

• **Standards for certification of science specialists in elementary schools and of life-science teachers in middle and junior high schools need to be developed. Ideally, however, the task should be addressed as part of a larger effort to define the flow of scientific education from kindergarten through high school.**

• **Efforts should be made to clarify the role of mentor teachers that are to be required for national board certification. Mentoring is not a common working concept for today's teachers, who perform their**

duties in isolation from one another and with great autonomy. Few could define a mentor or describe what one does. Mentor relationships between experienced and novice teachers, however, are a cornerstone of the reform movement. The subject of mentoring is discussed at length in the following section.

INSERVICE EDUCATION: HOW TEACHERS CONTINUE TO LEARN

Background

"Inservice education" refers to the formal, usually structured activities of practicing teachers intended to improve their knowledge or skills. It is a form of continuing education. Local districts have a major influence on inservice education, in that they can specify the kinds of academic credits or other activities that will advance a teacher on the salary scale. States, too, can require kinds and numbers of inservice activities for renewal of teaching licenses. More recently, professional teachers' groups have identified inservice education as a sphere they want to influence. Teacher associations, unions, and interested commissions and foundations have begun to recommend standards for inservice education (Green, 1987).

The nature of inservice activities is changing. About half the states require a master's degree for a professional license, which allows a teacher to teach virtually permanently. Today, 63% of all grade 10-12 science teachers already have master's degrees, so fewer teachers are returning to universities for further education. But to advance on the salary scale, teachers need continuing education. Recently, teachers unions have argued that teachers, rather than institutions of higher education, should provide, or at least select, the additional experiences or course work needed for professional renewal. Perhaps as a result, there has been a proliferation of short, topical workshops, usually offered by private educational consulting firms, at which teachers earn credit renewal units (CRUs) or continuing education units (CEUs). Educational consulting agencies charge for their work and are competitive; that can lead to reductions in time requirements and rigor to secure a contract. Such inservice activity is contracted by school districts, many of which have an administrator designated to select and organize inservice programs, or by local teachers unions. Although some skills can be taught by one practitioner to another, others, such as updating of content, require college researchers and teachers.

More familiar to scientists is the major attempt to reform and revitalize inservice education for science teachers that was undertaken by the federal government during the 1960s and 1970s. Academic-year institutes and summer institutes were sponsored by the National Science Foundation (NSF). Those programs had the following characteristics:

- They emphasized knowledge of the subject (e.g., biology).
- They introduced new curricular programs.
- They reflected the nature of the scientific enterprise and what scientists do.
- They involved formal course work and academic credits leading to a master's degree in biology.
- They paid teachers stipends and travel expenses for attending.

The primary focus of the institutes was on updating teachers' knowledge of science, and in this they were undoubtedly successful. They paid less attention, however, to changing teaching methods (i.e., to content-specific pedagogy).

A 1977 survey indicated that nearly 80% of mathematics and science supervisors and 47% of science teachers in grades 10-12 had attended NSF inservice institutes. However, only about 5% of grade K-3 teachers had attended such programs (OTA, 1988). Although the General Accounting Office (GAO) reported little effect of the institutes on student achievement scores (GAO, 1984), results of studies by the Congressional Research Service and the National Association for Research in Science Teaching indicate that the institutes had positive effects (OTA, 1988). Part of the confusion stems from the fact that the examinations that test achievement in biology were not well matched to the laboratory-based experiences offered by the institutes. Furthermore, it became clear to this committee through meeting with several hundred biology teachers that the NSF-sponsored summer institutes had a deeply positive effect on teachers' morale and sense of belonging to the wider scientific community. Critics of the institutes have dismissed that result as marginally important (GAO, 1984). But, in a profession to which it is difficult to attract talented people and in which it is a struggle to prevent burnout, inservice programs that help teachers to "feel good" about what they are doing certainly must touch the classroom in subtle, positive, and important ways. As we discussed in Chapter 4, average achievement-test scores tell us little about the understanding of science that students take from the classroom. In addition, they tell us nothing about how students are inspired by enthusiastic teachers to continue study of the subject in college.

The type of inservice programs available to biology and natural-science teachers has changed drastically since the 1970s, when both summer and academic-year workshops and programs were offered on many college campuses. Today, college inservice programs are offered for short periods, are "one-shot" efforts, and usually are not conceptually organized.

Not surprisingly, inservice activities in science are elected more frequently by teachers in the upper grades. In a 1985-1986 survey, only 16% of grade K-6 teachers, but 47% of teachers in grades 7-9 and 46% of teachers in grades 9-12, had taken a science course since 1983 (Weiss, 1987). (The percentage of teachers in grades 7-9 reflects an NSF program directive to enhance middle-school science.) Similarly, when teachers are asked how much time they spent on inservice education in science in the preceding year, half the grade K-6 teachers, but fewer than one-third of grade 7-12 teachers respond none. Discrepancies exist in the availability of inservice science education for teachers at various grade levels. Furthermore, NSF awards (in contrast with formula grants) are made to individual institutions on the basis of merit, so availability varies considerably with geographical area.

When Weiss (1987) asked teachers to describe their preferences for scheduling inservice activities, 60% responded that they would "very likely" attend if a workshop were offered on a workday. However, only one in three would be very likely to attend a summer or after-school meeting, and only one in five secondary-school science and mathematics teachers and one in seven elementary-school teachers would be very likely to attend Saturday or evening inservice programs. Yet almost all the earlier NSF institutes were on Saturdays, after school, or in the summer. Clearly, both the scheduling and substance of teacher inservice programs have changed dramatically in the last 2 decades. A lack of stimulating inservice activities for science teachers could be a major contributing factor in that change; however, if teachers' attitudes about the professional importance of keeping abreast of advances in science have changed, the problem is even deeper.

Reform Movements and Inservice Programs

After a period of severe budgetary cutbacks and elimination of programs in education, the NSF Directorate for Science and Engineering Education once again has funds in its teacher-enhancement and network programs, but there is no apparent cohesive strategy for improving the science teaching force. Although a plan for dissemination must be part of every proposal submitted to NSF, principal investigators have autonomy to develop individual projects and programs. NSF has not targeted for broad study and impact any specific components of inservice education, such as courses in new biological topics, improvement of students' quantitative abilities, development of mentor-teacher skills, integration of mathematics and science, or teaching to all students. In short, NSF has not developed a long-term agenda for reform. Disparate programs, delivered in a variety of ways, are not likely to address the substantive needs of practicing biology teachers nationally, nor to foster a professional camaraderie, as did the NSF inservice programs during the 1960s and 1970s.

Some inservice programs have nevertheless been successful (see Appendix E). But such programs for high-school teachers are too few and too limited in scope to address the national problem. The most successful examples should be examined to identify the elements that make them successful, and funds should be found to replicate these models so as to make outstanding inservice opportunities much more widely available.

Both the Holmes- and Carnegie-sponsored initiatives propose changing the nature of the teaching force; both advocate differentiated teaching staffs with professional or lead or mentor teachers (all terms are used) having more responsibilities and receiving more remuneration. For example, a Carnegie report (1986) recommends introducing "lead teachers" who can help to re-design schools and assist their peers in upholding high standards. Furthermore, certification by the National Board for Professional Teaching Standards rests on the identification and cooperation of existing teachers to serve as mentors. The members of the Holmes Group advocate that teachers in their professional development schools have high status and participate continually in giving and receiving inservice education.

None of those recommendations, however, focuses on the specific nature of inservice education that will change how science is taught in the nation's schools. Indeed, the Holmes Group assumes that teachers who have better preservice preparation will automatically continue in their field, whereas the National Board for Professional Teaching Standards addresses inservice education only as part of a 3-year induction process. Neither the members of the formal reform movement nor the informal task forces, consortia, or commissions have addressed the critical needs in inservice education for the approximately 37,000 biology teachers now in the profession, and all plans for reform require changes in preservice preparation that are far from being in place.

The Concept of Mentors

The use of mentors is not a working concept in today's schools. In other professions, a mentor is a trusted counselor who provides support and guidance to junior professionals. In our nation's schools, however, there are few documented examples of young teachers' benefiting from the knowledge and experience of master teachers. We consider interactions with experienced faculty an important part of the induction process, but young teachers are now expected to go out on their own after graduating from college, and only rarely are master teachers used to train them. When such interactions do occur, they usually take the form of reporting on skills or techniques learned during a workshop offered at a conference or summer inservice activity. Although those activities are desirable, a much greater potential exists for experienced teachers to contribute to the continuing inservice education of less-experienced teachers.

We also consider mentoring to be an important part of the professional development of master teachers. By assisting in the training of young teachers, older teachers can share valuable skills acquired through years of classroom experience with the next generation of teachers. Exemplary teachers should also be encouraged to devise new pedagogical techniques.

The Role of Mentors

Mentor teachers should play an important part in the continuing education of both novice and experienced teachers. They should give special attention to teachers during the critical first few years. In particular, mentor teachers would:

- Help novice science teachers to make the best use of class time, curricular materials, laboratory time, and resources.
- Help novice science teachers with practical guidance on instructional techniques.
- Support beginning teachers by providing ideas for science activities and information about local informal science resources.
- Notify beginning teachers of pertinent and quality inservice activities.
- Supervise novice teachers in both classroom and laboratory settings.
- Provide advice about school and district policies.
- Substitute for novice teachers, so that they can visit classrooms of more experienced teachers.
- Help experienced teachers change their teaching methods, enhance their laboratory programs, and improve their program of assessment.
- Encourage other science teachers to observe their teaching.

Identification of Mentors

There are few examples of mentor-teacher programs. Most teachers who serve as advisors of younger teachers are self-selected, because few, if any, institutional benefits are related to mentoring in its present context. Mentoring will not become a working concept without some drastic changes in how master teachers are used within schools. Several options are available to identify mentor teachers, but they are untested.

A mechanism must be developed, perhaps in the form of professional certification, to identify mentor teachers. The National Science Teachers Association (NSTA) certifies teachers who submit proof of having achieved NSTA standards for science teaching. Relatively few teachers, however, have chosen to become certified this way.

The National Board for Professional Teaching Standards potentially offers another type of certification for mentors. Connecticut's Beginning Educator Support and Training (BEST) Program (1988a, 1988b, 1988c) is another example of a new program that highlights the continuum of professional development from novice to mentor teacher; more efforts in this direction are encouraged.

Benefits to the Mentors

The goals of mentor teachers will be to provide advice and guidance to novice teachers and to help to institute curricular reform. Through participation in a mentoring program, experienced teachers would benefit by gaining the opportunity to teach other teachers at a higher level. The break in routine would be a rejuvenating experience for committed teachers who might otherwise suffer from "burnout."

To attract the best teachers into mentoring programs, incentives must be built into the system:

- Released time for mentor teachers to work with younger teachers should be funded.
- Mentor teachers should be scheduled to teach fewer classes per day and should use this released time to supervise new teachers.
- Sabbatical time should be given to teachers to work with university researchers to improve their laboratory skills and content knowledge or to work with science educators on new curriculum projects and inservice activities. They would take their knowledge and skills back to their home schools and conduct inservice activities for other teachers.

Conclusions

When biology teachers have inservice opportunities in the discipline, it is usually under the auspices of a local university that has received an NSF grant or occasionally a Department of Education Title II award for teacher enhancement. Those programs run the gamut—2-day workshops, summer research experiences in industry or at a university, long-term involvement in curricular development, and so forth. Because of restrictions imposed by universities or NSF, few of the programs offer academic credit to the participating teachers, and support for travel and stipends has been reintroduced only recently. Few are incorporated into a conceptual approach that leads either to an advanced degree or to a deep understanding of the discipline. National leadership is clearly needed in identifying and defining the kinds of inservice programs that will be most successful in fostering inquiry-based learning by students, in integrating biological information and content pedagogy effectively for teachers, and in generating mechanisms by which pedagogical skills can be propagated through the teaching profession. Reinforcement of the profession with mentor teachers who can contribute to the professional development of younger colleagues is also needed.

Recommendations

- **Science has a continuously changing frontier, and society is characterized by change generated in part by science and technology. Moreover, science teaching is a profession, and a notable difference between professions and other occupations is that professionals are responsible for their own continuing education. As part of their professional development,** teachers must engage in inservice activities that advance their scientific knowledge.

   • New effective inservice programs should be created. They should be:

—Attractive enough to induce many teachers to participate and appropriate to teachers' needs, as identified by teachers.

—Conceptually organized, eventually operate in conjunction with preservice programs, and run on a continuing basis.

—Able to compensate teachers for their time.

—Associated with opportunities for teachers to obtain small grants that enable them to bring new approaches to the classroom.

—Constantly evaluated for effectiveness.

—Scheduled with sufficient flexibility to ensure attendance.

—Designed to combine understanding of *what* to teach with understanding of *how* to teach.

—Designed and conducted with the collaboration of experienced science teachers, educators, and research scientists.

—Coupled to mechanisms for disseminating new information throughout the school district.

   • The assumption that summer institutes like those sponsored by NSF a generation ago, which gave mathematics and science teachers more up-to-date knowledge of their subjects, will necessarily lead to better teaching is naive. (As described in Chapter 8, the loss of interest in Biological Sciences Curriculum Study materials provides a lesson about the design of inservice programs.) In addition to teaching content, new inservice programs should be meticulous in developing an array of effective pedagogical techniques that engage students in learning scientific concepts instead of scientific jargon. Different models need to be tried and evaluated in novel ways. The effects on student performance should be recorded and the results reviewed by instructors and participants and fed back into program development.

   • The changing climate in which teachers operate suggests that some experimentation will be necessary to schedule inservice programs that will attract teachers, but also will be interesting, demanding, and rigorous enough to change how teachers operate in the classroom. Moreover, precollege science faculty should have flexibility to choose inservice programs according to the needs of individual schools and teachers and not have inservice options thrust on them by administrators who are not engaged in teaching.

   • Change for teachers will be gradual and will depend on their own perceptions, as well as student, parent, and community perceptions of improved results. Teachers therefore need support after the inservice work (Guskey, 1986). They must be given time to assimilate the knowledge and suggestions proposed during inservice programs, to consider how the changes will affect teaching and learning in their classrooms, and to consult with colleagues. Attention should therefore be given to the need for long-term collaborative arrangements for inservice support. Such inservice activities should focus on collaborations among industry, the university research community, and schools. They should involve federal and private sponsorship, rather than be expressly commercial. And they should be guided by the proper educational criteria, as described above.

• What has been said about inservice programs for high-school teachers generally holds for elementary-school and middle-school teachers. The available models are fewer, and the backgrounds, motivations, and interests of the teachers are different, even though the challenge is no less important. Because the nature of the task is different from that posed in the case of high-school teachers, we recommend the development of distinct cooperative inservice programs for elementary-school and middle-school teachers.

• Some partnerships between school districts on one side and universities, foundations, and local industries on the other are trying to enhance classroom performance. As suggested by the preceding two recommendations, more need to be tried. Owing to local autonomy, the difficulty with this approach in isolation is that it will tell us little about what works or why. We need improved mechanisms for assessing the success of the various experiments in inservice education that are under way. The assessments must be more sophisticated than the traditional recourse to average scores on regional or national examinations. Pedagogical skills can be identified, taught, and assessed (Kahle, 1985, 1987). We also need improved methods for distributing information developed in successful inservice programs, and we need to attract the participation of additional teachers. These are not matters that can be accomplished in a single effort; continuous evaluation involving longitudinal and case studies will be required. In Chapter 8, we suggest how the research community might collaborate with teachers and others to accomplish these goals.

• New inservice programs should address the need to develop a larger cadre of mentor teachers. Graduates of such activities should be prepared to provide assistance to other teachers in their schools and districts. Fellowship support, perhaps even in the form of sabbaticals, should be made available to the best teachers, so that they can go to universities, improve their skills and subject knowledge, engage in science or education research, participate in curriculum development, and prepare themselves to assist less-experienced colleagues when they return to their home schools. Every effort must be made to keep such teachers in the classroom, including salary structures that reward, rather than penalize, talented teachers for remaining teachers.

6

Recruiting Scientists, Teachers, Technicians, and Physicians

THE GENERALITY OF BASIC EDUCATIONAL GOALS IN SCIENCE

The efforts to improve biological education that occurred in the 1960s were motivated by the competition of the Cold War and an anticipated shortage of scientists and technicians. The present concerns with science education, however, run deeper. There is good cause to worry about the rate at which we are training scientists, mathematicians, and engineers, although the shortages are more severe in some fields than in others (Vetter, 1987; Atkinson, 1989; Green, 1989). We are also concerned that an insufficient number of talented and energetic young people are choosing to pursue careers as precollege science teachers. In addition, the limited interest in and understanding of science by most students, many of whom do not pursue further schooling after high school, constitutes a serious national problem. These students will become voters in a society of increasing technological complexity. Increasingly, they will seek employment in technical settings in which capacity for observation, inference, and simple calculation are likely to be prerequisites for even entry-level positions. In many ways, therefore, the students who do not continue with science constitute our largest challenge.

We are convinced that instruction in biology that plants the seeds of discovery, awakens students to the beauty of the world around them, and instills some understanding of fundamental biological concepts will serve the interests of most students. Many students who now enter college expressing an interest in science nevertheless look back on their high-school science experiences as obstacles they have surmounted, rather than as gardens they have entered. The basic goals of a good tenth-grade course are appropriate for all students—those who will become scientists and those who will pursue other careers.

The phrases "science as a way of knowing" and "science as process" carry the conviction that science should be learned by doing. A substantial consensus has developed among investigators of "giftedness" that an environment that encourages inquiry provides the best opportunities for all students to learn (Brandwein and Passow, 1989). The role of the laboratory (as described in Chapter 4) is therefore central to successful instruction; if opportunities are made available to all, students with the appropriate abilities and interests will identify themselves with scientific activities with an appropriate degree of challenge (Brandwein and Passow, 1989). In some schools, it might be possible to provide opportunities for involvement in the scientific process outside the classroom and outside the curriculum. That involvement can be especially important in sustaining the enthusiasm of the students most likely to pursue careers in science.

AN IMPORTANT ROLE FOR UNIVERSITIES AND UNIVERSITY SCIENTISTS

We have not been very systematic about our quest to improve teaching, even though we value it highly and frequently do well at it. I am struck, for example, by the lack of conversation about what pedagogy means, and what makes it successful. It is our profession, yet it is mysteriously absent from our professional discourse. Here we are, engaged in an activity that is vital to ourselves, our students, and our public—yet we speak of how to do it, if at all, as though it had no data base, lacked a history, and offered no innovative challenges.

Donald Kennedy (1990)

Most universities in the United States are staffed by outstanding science faculties. The faculty members are grouped into a series of departments, which are formed according to traditional research disciplines, such as biology, biochemistry, chemistry, physics, and geology—each with between 10 and 60 faculty members, depending on the discipline and the size of the university. The world leadership of the United States in the sciences depends on the excellence of these departments, which is maintained by competitive forces of three types:

(1) Each faculty member competes nationally for the research grants that are required for the establishment and continuation of a first-class research effort. For biological scientists, the major funding sources are the National Institutes of Health (NIH) and the National Science Foundation (NSF). The competitive nature of the system guarantees that the limited resources available for biological research are distributed to those best able to use them.

(2) In each field of science, different universities compete to attract out-standing graduate students from a limited pool of college graduates. Success in recruiting these sophisticated young people nationally and internationally requires that the faculty members of each department work together to create

an effective graduate teaching and advising program, in which each member's efforts contribute to a departmental atmosphere in which most graduate students will prosper.

(3) The central university administration often apportions new faculty positions and other resources among the many departments in its university according to the total number of students that each serves. Each department in a university therefore competes with its sister departments to attract undergraduate majors. The courses that departments offer to undergraduate science majors are usually taken quite seriously by both the faculty and the department chairmen. If these courses are well taught, the department will be able to recruit a larger number of undergraduates to major in its discipline, and over the long run the department can hope to expand. Thus, competition for the limited pool of students within each university has a beneficial effect on the quality of the education that its undergraduates receive in their major fields.

The various competitive forces described above encourage faculty members to concentrate their efforts on their own laboratories and on graduate and undergraduate teaching programs that cater to the students wishing to specialize in their discipline. Most successful faculty members in science departments are already overcommitted from their efforts in research and "professional" teaching (the teaching of graduate students and undergraduate majors). It should therefore not be surprising to find that other important responsibilities of the university— in particular, providing good science training to precollege science teachers and to nonscience majors—have traditionally been seriously neglected.

At present, the major pressures with regard to the aspects of a science professor's job that are most relevant to this report are all negative ones. Because the time and effort spent on precollege teachers and on nonscience majors are bound to detract from the time available for department majors, graduate students, and research efforts, success in the former endeavors may be viewed as hurting the department, rather than helping it. Thus, the current reward system for university faculty helps to explain the unsatisfactory nature of the general science training that we provide to most of this country's undergraduates (Westheimer, 1987).

The current reward system also explains why most university scientists and science departments do nothing at all for the precollege science teacher. By providing neither preservice nor inservice training, most university science departments have cut off themselves and their students from the world of precollege education. This is clearly detrimental for the precollege teacher; in addition, the faculty thereby sends a clear signal of disapproval to any student who might otherwise wish to choose such a career.

University science departments therefore must accept a share of responsibility for the present unsatisfactory situation with regard to science education. Despite their world-renowned excellence in scientific research and advanced teaching (or perhaps because of it), our universities have become a significant part of the problem of inadequate K-12 education in science. The university scientists must strive to make precollege science teachers visible members who are respected by university scientists and treated as colleagues with shared goals. That represents a complete change from the present situation, in which precollege science teachers are routinely ignored by the research universities in their communities. Precollege teachers will benefit from the help and status that they derive from their university contacts. But the universities will also benefit from the interaction, because their students will become familiar with many precollege teachers and thereby with an important career option that most of them would otherwise never consider.

Recommendations

- **The various national organizations of college and university presidents, professors, and administrators should be made aware of their role in the crisis in precollege science education in the United States. They should act together, in concert with professional organizations of scientists, to develop new reward systems on each campus that counteract the present tendency of science departments to ignore the training of precollege teachers. In particular, university leaders must make it clear to their science departments that the quality and quantity of the service that each department provides to precollege science teachers (both preservice and inservice training) and in the general education of nonscience majors will be important considerations in the distribution of university resources and faculty positions.**
- **Universities should develop programs that integrate all interested local precollege science teachers into the various science communities of the university. Possibilities include partnership programs between the precollege teacher and faculty, postdoctoral researchers, technicians, and graduate students; periodic laboratory tours for teachers and their students; availability of surplus equipment and supplies to the teachers; science contests for students; and summer research opportunities for both teachers and outstanding students.**
- **Major universities should be encouraged to develop permanent summer inservice institutes for precollege science teachers, on the basis of successful model institutes held elsewhere. Outstanding faculty from university science departments should be recruited to teach in these institutes, side by side with outstanding precollege mentor teachers.**
- **The professional organizations of teachers, scientific societies, and state academies of science can do more to provide information to students on college and university campuses about careers in science teaching. For example, the National Association of Biology Teachers could work with**

college departments of biology and career-placement offices to schedule visits by outstanding precollege teachers and otherwise to make sure that information about opportunities in teaching is readily available.

THE NEED FOR A NATIONAL FELLOWSHIP PROGRAM TO ATTRACT OUTSTANDING YOUNG PEOPLE INTO TEACHING

Gifted students or those who are excited by biology do not require a conceptually different curriculum from other students. The needs of all students will be best met by seeing that classrooms are staffed with able teachers who have a deep understanding of concepts, enjoy teaching science as a process of discovery, and are flexible and creative in addressing the needs of individual students. All the recommendations in this report are directed to that goal. The future supply of outstanding science teachers is therefore central to improving science education.

Recruitment of future life-science and biology teachers (grades 6-12) and of those able to teach natural sciences (grades K-5) is a major issue with which none of the reforms suggested by the Holmes and Carnegie groups deals directly. Recommendations have been made to raise both the entrance (grade-point average) and exit (National Teacher Examination or other competence-examination score) requirements for programs educating teachers. Such changes might improve academic competence, but they will not address recruitment of future teachers. Common practices, both subtle and overt, discourage biology undergraduates from electing careers as precollege teachers.

Discouraging able students from teaching careers begins in the high schools, where biology teachers actively encourage their best students to become scientists, not science teachers (Kahle, 1985). It continues in college, where biology faculty discourage undergraduates indirectly by failing to present information about teaching options and directly by opining that precollege teaching is not a worthy choice of career for able students.

Recommendation

A particularly promising mechanism for addressing teacher recruitment is a national fellowship program in which selected prospective teachers are paid for their schooling. Such a program could have several features that would further collateral goals as well:

—A competitive fellowship program could attract some of the most able biology or education majors into science teaching.

—It could be of immense help in reaching minority groups that are underrepresented in the teaching force.

—With foundation underwriting, it would be possible to couple the use of fellowships with institutions that have shown interest and imagination in addressing the kinds of changes that are required in preservice education. Prestigious fellows who have studied at institutions that created the best conditions for research have made some of the most important contributions to science, and a corresponding formula needs to be tried in education.

—Fellowships need not be exclusively for future high-school science teachers; they could be used to attract future science specialists for elementary schools and middle schools, thereby boosting the prestige of that calling, as well as increasing the numbers of such teachers.

—Similar fellowships could be offered to established teachers. In the most ambitious form, they might be used to underwrite year-long sabbaticals, during which teachers would attend a university and participate in the development of new preservice and inservice programs, as well as improve their own knowledge of science.

SEX, DEMOGRAPHICS, AND RECRUITMENT

Most scientists, engineers, and science teachers have been white men; women and minority groups have been underrepresented. The asymmetry has many reasons, and there are many reasons for seeking change. Among the latter is one that follows from demographics. Forecasts for the United States indicate a shifting composition of the workforce, which will be aggravated by a projected shortfall of scientists and engineers in the coming years (Atkinson, 1989; Green, 1989). In 1982, the population of students 5-17 years old was 73.3% white, 14.9% black, and 8.9% Hispanic; but by the year 2020, these numbers will have become 52.7, 19.8, and 23.9% respectively. In 25 of the nation's largest cities, minority groups already contribute most of the students (Vetter, 1989), and the trend is expected to continue. Yet those fastest-growing segments of the American population are generally lost from the science pipeline. Women and minority groups must be encouraged to help to bridge the gap between the decreasing supply and the increasing demand for scientists and engineers.

Research suggests why women and minority-group members do not traditionally enter science and engineering careers. Among the cultural and social factors contributing to the male-female discrepancies in achievement in science and in choice of science as a career are educational and attitudinal differences between boys and girls that influence who goes on to study science and mathematics (Kahle and Lakes, 1983). A growing body of research indicates that many white girls and minority-group boys and girls have substantially different experiences in science from white boys in grades K-12. For example, they have fewer routine daily experiences with the tools, materials, and equipment of science, and they are called on less often in science classes (Kahle and Lakes, 1983; Whyte, 1985). Moreover, an analysis of 1976-1977 National Assessment of Educational Progress (NAEP) data showed that, "by age nine, females, although expressing similar or greater desires to participate in science activities, had consistently fewer experiences than boys of the same age. . . . At ages 13 and 17, girls again reported fewer classroom and extracurricular science activities than boys" (Kahle and Lakes, 1983, p. 131).

Similarly, a study of minority-group attitudes toward science based on 1976-1977 NAEP data showed that, "overall, black students appear to like science, but, perhaps due to their limited exposure and experiences, many do not understand its methodology, technology or potential" (Kahle, 1982, p. 542). The author concluded that science programs and curricula had not capitalized on the positive attitudes exhibited by members of minority groups.

Women and minority groups are often dissuaded from science, either consciously or unconsciously, by teachers, parents, and guidance counselors (Linn and Hyde, 1989). Moreover, because they often do not have as much exposure to science activities as white males, their ability to decide about careers in science and engineering is more limited. As the Committee on Policy for Racial Justice (1989, p. 2) has concluded:

> The essential problem lies not with the academic potential of black children but with unproductive institutional arrangements, lowered expectations, and narrow pedagogical processes that characterize the American educational system.

Attracting Women and Minority-Group Members into Research Careers in Biology

Numerous public and private efforts are being made to increase the participation of women and minority-group members in science and engineering careers. The Task Force on Women, Minorities, and the Handicapped in Science and Technology, established by Congress, was charged "to develop a long-range plan to advance opportunities for women, minorities and the handicapped" (Task Force on Women, Minorities, and the Handicapped in Science and Technology, 1989). The task force was composed of representatives of 15 federal agencies and leaders in the private sector and education. It has issued recommendations requesting that actions be taken by school boards, states, the federal government, universities, industry, and the entertainment media to create a climate of high expectation for students from underrepresented groups to pursue careers in science and engineering. Private foundations and institutes have also taken an active interest in encouraging women and minority-group members to pursue careers in science and engineering. Appendix F describes several of their efforts.

Attracting Women and Minority-Group Members into Teaching Careers in Biology

The increasing cultural and linguistic diversity of the student population needs to be matched by an increasingly representative teaching force. Apart from considerations of social equity, diversity constitutes a pragmatic reason for wishing to have better representation of minority groups among the nation's teachers. But minority-group teachers are leaving teaching, and the number of new ones entering the profession is decreasing, in large part because of the lure of other opportunities. In the view of some, the widespread use of standardized tests—both admission tests, such as the SAT and ACT, and teacher certification tests—has put minority groups at a disadvantage in entering the profession (Ryder, 1989). In addition, the enrollment of blacks in 4-year colleges and universities has decreased during the 1980s. Although Hispanic enrollment has increased since 1980, this group is still underrepresented (Alston, 1988; NRC, 1989b). Unless more minority-group students go to college, the pool from which minority-group teachers will be drawn will remain small.

Currently, minority groups (black or Hispanic) make up a small percentage of science teachers: 13% in grades K-3, 12% in grades 4-6, 7% in grades 7-9, and 6% in grades 10-12 (Weiss, 1987). Similarly, whereas 94% of teachers in grades K-3 and 76% in grades 4-6 are women (grades for which little or no specific science training is required), only 41% of science teachers in grades 7-9 and 31% in grades 10-12 are women (Weiss, 1987).

Conclusions

Studying science as a process of discovery about the natural world is appropriate for all students, whether or not they will ever be professionally engaged in science and teaching. Practices that discourage girls and minority-group children from achieving their full potential in science and mathematics must be identified and eliminated. The disparity between the number of minority-group children and the number of minority-group teachers is growing. Moreover, programs for educating teachers must compete with other attractive professions for students. Current efforts to restructure the teaching profession—which will result in greater professionalism, higher salaries, better working conditions, and more appropriate assessment techniques—will all make teaching more attractive to the ablest minority-group students.

Recommendations

- **At all grade levels, teaching strategies are needed that encourage the active involvement of girls. That will be easier if the teachers in the early grades, most of whom are women, can be made more comfortable with** science. The recommendations in the sections on preservice and inservice programs bear on this need.

• Those who develop preservice and inservice science and science-education courses need to be aware of research that demonstrates how prospective and practicing teachers can develop and implement specific teaching behaviors and instructional strategies that lead to more equitable science classrooms.

• Greater attention should be paid by colleges and universities to recruiting women and minority-group members to careers in science and science teaching. To that end, stronger links could be forged between the historically black colleges and graduate and professional schools in research universities. Community colleges represent another source of potential talent that has not been fully tapped.

• To interest students in science, and women and minority-group members in particular, teachers should be provided with up-to-date information on career opportunities in the biological sciences, including such diverse fields as biotechnology, agriculture, wildlife management, ecology, and biomedical research. Ideally, such information should be supplied as part of preservice education, but to keep up with changing technologies and demographics it should be updated regularly and made available nationally. The task of conveying the information to students requires special knowledge of biology and is therefore not appropriately left entirely to guidance counselors.

Much of the organization and support of our system of public education originates locally; constructive involvement of everyone with a stake in the outcome is essential to its success. For example, early awareness of the wide array of vocational opportunities available to students with an interest in biology should help motivate them and raise their expectations for success after graduation. The information provided to students must be current, must relate to local conditions, and must, wherever possible, engage the cooperation of potential employers. The participation of local employers can vary widely: making known their needs for entry-level positions, involving themselves directly in inservice programs, offering advice to students and teachers on research projects they have undertaken, and providing opportunities for students and teachers to become engaged in research projects. The various community-based activities that foster involvement of parents (such as some of those described in Appendix F) should be encouraged and extended.

7

Other Modes and Contexts
for Teaching Science

INTEGRATING BIOLOGY WITH OTHER SCIENCES

In Chapter 3, we assumed that the high-school science curriculum in most schools in the immediate future will retain the structure it has now: it will include biology as a separate course, usually the first science course taken in high school. (It usually comes after a variable and generally inadequate exposure to "health" and general science in middle school, but this can and must change.) A high-school science curriculum that proceeds from biology to chemistry to physics entails some substantial educational problems. We touched on some of the issues in Chapter 3, but we take them up again here, because an adequate solution will require a long-term approach.

The fundamental problem arises from the increasing need to understand some chemistry in order to understand much biology. Students now enter high-school biology knowing little or no chemistry and physics. The pragmatic "solution" has been either to teach aspects of chemistry in the biology course, to require students to memorize biochemical names and organic chemical structures in a context destined to kill interest, or to combine the two. If students have not even studied enough chemistry to know that "carbon has a valence of 4" or even to comprehend what that statement means, there is no justification for expecting them to know the much more complicated molecular structures of glucose and alanine.

Suppose the sequence of courses were reversed, with physics preceding chemistry and biology coming last. That arrangement also has its difficulties, in that physics and chemistry are more successfully taught to students who have more mature capacities for abstract reasoning and more extensive experience with mathematics.

The solutions to the dilemma are educationally interesting. If children in elementary school were to have a steady involvement in science with an emphasis on natural history, as proposed in Chapter 3, they would enter high school more aware of the world around them, the diversity of life, the relationships among living things, and the structure of their earth, the atmosphere, and the planets and stars. Students coming from such a background would already be aware of fossils, for example, and would step naturally into a study of how evolution can be inferred from the fossil record. They would already know enough about plants and animals to appreciate the differences among most of the commonly observed taxa. Students would be exposed to integrated subject matter *before* they entered high-school biology, and they would have a far better base for learning biology than most students now have when they start the subject.

But that is just the start; another kind of integration of subject matter could occur during middle school and high school. In almost every developed nation—but not the United States—secondary schools teach biology, chemistry, physics, and mathematics either in parallel streams or in integrated multiyear courses. The pedagogical advantages of those approaches are clear and obvious. The entire problem of how or when or whether to teach the more molecular aspects of biology would disappear, because the necessary chemistry could be presented before the corresponding part of the biology curriculum is reached. All the subjects would be integrated as soon as the relevant bits were presented. The sense that the sciences truly constitute a unified, integrated body of knowledge would no longer depend on whether (and when) a student could fit 4 years of separate packages into a conceptual whole. The student who took only 1 or 2 years of high-school science would learn some of the most basic and important concepts of all three disciplines—biology, chemistry, and physics—instead of missing one or two of the disciplines almost totally.

One drawback of having integrated or parallel courses is that it would require much work to prepare them and much more cooperation among teachers than does the present system of separate sequential courses. In the view of this committee, that is an insufficient reason for not developing an integrated or parallel science program. The benefits in scientific literacy, in coherent and logical presentation of subject matter, and in arranging the subject matter to fit students' developing conceptual sophistication far outweigh the short-term difficulties of redesigning a curriculum.

Recommendations

We should begin now to plan and support models for integrated or parallel programs in biology, chemistry, physics, and mathematics, both for high schools and for grades K-12. The details of the curriculum might turn out to be the easiest part of the task, because to effect the change on a broad scale will require in the short term the creation of appropriate inservice support, the development of new patterns of cooperative teaching, and the cooperation of teachers. The new National Science Teachers Association Scope, Sequence and Coordination project (Aldridge, 1989) is a move in

that direction. In the longer term, preservice programs will have to be altered, as will expectations for licensing and certification of teachers. But the disadvantages of compartmentalizing the natural sciences at the high-school level will only worsen with time. We therefore propose as a first step a study of the benefits that might accrue from such a change in the curriculum and an analysis of the inherent obstacles to the implementation of this change. There will be wide-ranging implications for university curricula and a need to engage college and university faculties, as well as teachers, in these reforms. Chapter 8 presents a possible mechanism to help advance our recommendation.

ADVANCED-PLACEMENT BIOLOGY

The Present Advanced-Placement Program in Biology

The Advanced-Placement (AP) program in biology, which is sponsored by The College Entrance Examination Board (College Board), consists of a course description with suggested time percentages for major and minor topics, suggested textbooks, laboratory exercises, an examination, and a list of more than 2,000 colleges and universities that "*normally* use Advanced Placement Examination grades in determination of advanced placement and credit" in biology (College Board, 1987, p. 77).

In principle, the AP biology course is intended to be equivalent to an introductory college-level course in biology. To plan the most recent revision of the course, 80 colleges were surveyed in 1985 to determine the content of their introductory courses for biology majors. The AP course was designed on the basis of the results of the survey. Three major sections are outlined: molecules and cells (25% of allocated time), genetics and evolution (25%), and organisms and populations (50%). Each section is divided into topics (with suggested allocations of time). The design of the course is predicated on the assumption that students have successfully completed year courses each in high-school biology and high-school chemistry.

Laboratory work in the AP biology course is based on data that suggest that about one-fourth of the credit for college biology is derived from laboratory work. The course guide presents 12 laboratory activities, all experimental and quantitative, with detailed advice for all aspects of each activity. AP biology teachers are expected to integrate those activities into their curricula and to conduct additional laboratory activities. The laboratory activities are considered to be "basic introductions, or springboards, into further experiments, studies, or independent projects" (College Board, 1987, p. 7).

First-level college biology courses typically consist of 40-50 hours of lecture and 25-30 hours of laboratory work per semester, and equivalent time should be allocated for the AP biology course. School administrators and prospective AP biology teachers are warned of that requirement and warned that the AP biology course, if it is to be equivalent to a college-level course, will be substantially more expensive than a typical high-school biology course.

The AP biology examination consists of a 90-minute, 120-item multiple-choice section and a 90-minute section of "free responses" or essays based on four mandatory questions. There is one question each for the first two major content sections and two for the third. To ensure that laboratories are used in AP biology classes, some questions are related to the laboratory experiments. The examination is designed to have a mean score of about 50%. Teachers are asked not to prepare students to answer every question, but to teach for understanding of the concepts. The rationale is that students who understand on a conceptual level what they have studied will do better on the test. The examination is graded by more than 1,000 college and secondary-school teachers familiar with the AP program. The multiple-choice sections are scored with a correction factor to compensate for guessing. In 1988, 64% of students who took the AP biology examination earned scores of 3 or higher (MacDonald, 1989)—grades deemed high enough to qualify for college credit or advanced placement in many (but not all) colleges and universities that recognize AP courses.

Over the decade 1978-1988, enrollments in AP biology increased from about 11,000 to 31,000. MacDonald (1989) states that AP students perform in college as well as or (often) better than non-AP students taking the college-level course for which AP credit was sought and tend to demonstrate higher achievement than their non-AP counterparts. That is not surprising, considering the goals and motivation of most students who take both the AP biology course and the examination. There is also a "good correlation between scores on the AP biology examination and subsequent grades in introductory and upper-level biology courses in college" (MacDonald, 1989)—again, not surprising or particularly revealing. Students often report that they found themselves well-prepared for the sequence of advanced college-level courses in which they could enroll, but that view is not universally shared by college faculty.

The Success of AP Biology

If the recommendations of the College Board are followed by a properly prepared teacher with adequate laboratory facilities, the AP biology program could provide the equivalent of an introductory college biology course. The course has recently incorporated experimental, quantitative laboratory activities as an integral part of the curriculum. Compared with other commonly used assessment instruments, the AP-biology examination questions are much more advanced in their reading level and more effective in assessing the major ideas of the course and the general quality of understanding of the students.

The presence of AP biology provides an incentive for students with an interest in science and might serve as a device to recruit students to other science courses. And AP courses probably also help individual students in admission to college. It has also been an incentive for teachers who are willing to put in the extra work for an AP course in return for having a small group of motivated students. As an alternative to teaching in the common core curriculum, AP courses offer teachers some of the advantages of teaching a homogeneous group of motivated students.

Opinions of Teachers and Parents

Although there have been no extensive studies of the reaction of teachers and parents to AP courses, concerns that are voiced by parents, students, teachers, counselors, and administrators influence decisions to install AP science courses. For example, parent advocates argue that AP courses are more challenging than existing science courses and therefore more likely to motivate their children. They also argue that having their children take AP courses lowers their tuition costs (a spurious argument, except for students who graduate from college in less than 4 years) and that non-AP students in college classes with students who have taken AP courses are at a competitive disadvantage. They feel that it is appropriate for public schools to offer college-level classes for high-school students who can benefit from them.

However, serious problems, both philosophical and practical, attend the AP biology program. Some teachers feel that AP courses require more preparation time and more laboratory equipment, that textbooks (which are provided to students) cost more, and that students take a second year of biology in place of other valuable science courses that are available. And high schools are often not able to provide the resources necessary for a college-level course. Many biology teachers report that their school districts do not or cannot support the kinds of laboratory activities and field trips considered desirable for even the *regular* biology courses.* Some teachers report that the AP course covers too many aspects of biology in too short a time, puts excessive emphasis on lecturing by the teachers, does not devote enough time to laboratory work, requires teaching to the examination, and induces some of the most academically able students to take a course merely to gain admission to college. Other teachers, however, feel that AP courses influence students to take more rigorous academic programs. Counselors feel that there are valuable, rigorous non-AP courses that students reject in favor of AP courses. Administrators are concerned with taking on college-level responsibilities, with the costs of college texts and laboratory materials, with personnel problems (AP teachers' teaching loads can usually not be reduced or their preparation time increased), and with the impact on other course enrollments.

We are concerned that the AP biology course has been modeled on introductory college biology courses that for many students are notoriously poor educational experiences. The time has come to stop designing curricula by the process of serial dilution, in which the high-school course is a thin version of the college course, and the middle-school course is a thin version of the high-school course. The question of how well the AP biology course prepares students for upper-level biology courses is difficult to answer. There are no comparative assessments of how well college introductory biology courses prepare students. Moreover, many colleges and universities do not exempt students

---

*Biology teachers reported this state of affairs to the Committee on High-School Biology Education during a symposium at the NABT 50th Anniversary Convention, November 17, 1988, Chicago, Illinois.

with AP credit from their introductory biology courses, and others do so with misgivings. In some cases, students who have taken AP biology and passed the examination with a high grade are allowed into honors sections in college introductory biology; but it is not known how widespread or valuable this practice is.

Another matter, although of less concern, is that some students who take the AP biology course do not take the examination. The extent of that practice and the reasons for it are not clear, but its impact on the examination scores might be significant. The statement that 64% of students achieve grades of 3 or higher (MacDonald, 1989) obviously refers to those who take the examination, not to all those who take the course. The requirement of payment (by the students or the school system) for the examination might be a factor in decisions not to take it.

Critics feel that the AP biology course in particular and AP courses in general might contribute to tracking, can become elitist, and can compromise equity. Our committee, however, does not feel that offering advanced courses to interested students should become an issue of equity. The major question should be whether the courses accomplish their goals.

Conclusions

Secondary schools need to provide opportunities for able students to become passionate about their interests, whether in art, music, sports, humanities, or the sciences. We do not question the desirability of second-year biology, only the nature of the existing AP course. The present version of the AP biology course can have the positive effect of providing second-level opportunities for motivated students to study the science. In a number of cases, AP biology has doubtless provided opportunities for teachers and students to extend their knowledge and engage in exceptional educational experiences. We are skeptical, however, about whether AP biology is commonly able to provide an exposure equivalent to that offered in most colleges.

Recommendations

- **A consensus needs to be reached as to what the AP biology course should be. The present policy of modeling the AP course after a composite view of college courses is missing opportunities for generating a unique high-school experience, providing a more realistic introduction to experimentation, and providing better college preparation. Although the recent inclusion of quantitative experimentation in the AP program was needed and is commendable, an introductory college course may not be the soundest educational experience for students who have time for a second course in biology in high school. Whether the AP course will develop into a strong component of biology education or will itself become an obstacle to reform is unclear. A national body of educators, high-school and college biology** teachers, and scientists should make specific recommendations about the AP curriculum, examinations, and college credit. (See also Chapter 8.) The College Board should be asked to study fully its own record of success, follow up on the college placement of students, and assess compliance of high schools with its recommendations for prerequisites.

• Whatever their form, AP or other advanced biology courses should not be taken instead of chemistry, physics, or mathematics. Nor should they become the "honors" section, taken in lieu of the first high-school course in biology. The AP biology course should be taken as late in high school as possible, preferably in the senior year, to enable the subject to be taught as an experimental science to students whose maturity is close to that of college freshmen. Even a properly designed AP course in biology is inappropriate for younger students and for those without maximal preparation in mathematics and the physical sciences.

• We suggest that the terminal-year AP biology course provide intensive treatment of a few topics in molecular biology, cell biology, physiology, evolution, and ecology. Emphasis should be on experimental design, experimentation and observation, data analysis, and critical reading. Thus, the course cannot be modeled after existing college courses, which broadly cover all biology. Engaging students in direct investigations of natural phenomena without attempting to "cover" the subject matter of the introductory college biology course is judged by this committee to be more educationally sound than the current program. A rigorous examination devoted to problem-solving that requires the application of biological concepts should accompany such a revision.

• This course should be taught only by teachers both capable of providing a sophisticated and broad knowledge of biology and having the ability, training, experience, resources, and time to oversee an independent experimental approach. For example, a teacher who has not had first-hand experience in independent research should not be assigned to teach AP biology. Specific inservice training and certification should be used to ensure that only qualified teachers teach the AP course. The College Board should take initiatives to see that the program meets those more demanding specifications, but school administrators must understand and cooperate as well. If AP science courses are to be offered, there should be a line item in the school budget to support them, and they should not be given at the expense of regular science laboratory activities.

• The premise that AP courses provide college credit is not necessarily flawed; however, the nature of what the credit is for needs examination. A second course giving instruction in scientific reasoning, based on experimental design, and using sophisticated physical, chemical, and mathematical, as well as biological, principles would in fact provide better preparation for college than the present broad review. Colleges and high schools should both recognize the value of a second course in *experimental* science taken at the end of high school. Such a course need not be sponsored by the College Board or be designated "advanced placement."

A CAPSTONE HIGH-SCHOOL COURSE IN SCIENCE

Rationale: Integrating Science and Society

After leaving high school, many persons become public officials, civic leaders, corporate officers, or holders of other positions who must reach conclusions on issues that have scientific content and require integration of multidisciplinary information. Moreover, all students become eligible to vote, and the general public needs greater understanding of how different kinds of information are related to societal problems. Courses in specific sciences or other disciplines are unable, by themselves, to provide appropriate experience in integrating information from disparate sources. Furthermore, entry-level courses do not provide appropriate depth of laboratory, library, and community experience to generate and assess such information. We are concerned that courses offered as "science, technology, and society" (STS) usually do not follow a study of the basic sciences. Instead, they typically replace basic-science courses, and that results in both a dilution of fundamental knowledge of basic sciences and a lack of the scientific breadth needed to study interdisciplinary topics more than superficially. Although in the teaching of basic science new facts and concepts must be related to the learner's understanding of the world, we see danger in formats that confuse scientific knowledge with political, economic, and moral judgment. The contribution of science to the solution of societal problems can be understood only when there is considerable understanding of science itself.

We propose that an interdisciplinary "capstone" course be offered in the last year of high school, after students have already taken courses in biology and the physical sciences. The course would consider examples of current, major scientific technological-societal problems. It should use an integration of scientific, social, ethical, economic, political, and other disciplines to reach conclusions. Such a course would not be a simple extension of the science courses taken, but instead would focus on the integration of biological and physical sciences with the humanities and social sciences through consideration of contemporary problems. Such a course should not be substituted for chemistry and physics.

Organization and Content

The capstone course could be offered as a series of projects and could be taught by a team of teachers with particular interests in the individual topics relevant to the projects. The specific topics could be selected on the basis of the teachers' interests and expertise. The lead teacher should have advanced training in science. Examples of topics that could be included in the course today are acid rain; agricultural biotechnology; human applications of biotechnology; toxic wastes and pollution of groundwater; technology and development of the less-developed countries; environmental values; nuclear energy, fossil fuels, renewable energy, and commercial power requirements; and the ecological, sociological, and economic impacts of population growth.

The topics used need not have simple or even scientific answers. Students should define an issue, delineate the scope of the problem, and discuss the range of possible solutions, as well as the limits of available information. During analysis of the topic, students should debate the pros and cons, and teachers should not provide "answers." Thus, students will encounter the complexities of science and society firsthand and will recognize that simple answers are rarely possible or appropriate. The outcome of each project should include a comprehensive report written by each student that presents a description of the problem, alternative approaches and hypotheses, available data, and conclusions and recommendations. The writing component is essential: it not only ensures integration of information, but also requires the student to express analyses and conclusions clearly and concisely.

The capstone experience is not intended as an advanced-placement course. It should provide increased depth and breadth of knowledge in science and other disciplines and experience in weighing different kinds of information in making decisions. But it should be considered a course in science.

Benefits and Costs

The primary benefit of the capstone course is the educational reward to students in discovering interdependences, complexities, dilemmas, ambiguities, and the need to synthesize information in designing solutions to society's problems. Such a course will develop skills in reading critically and will foster understanding that scientific inquiry is open-ended and that studying science is not simply reading and memorizing. Where appropriate resources are available, the capstone course can facilitate the use of technology in the analysis of data (e.g., use of computers to analyze data both graphically and numerically), as well as provide direct experience in conducting literature searches. It can also allow the development of relationships with resource persons and agencies in the community and provide new mechanisms for teachers to participate in continuing inservice training and development.

A capstone course cannot be implemented without incurring substantial costs and difficulties or without rethinking of teaching practices. Incremental resources obviously will be needed to develop and test curricula, buy equipment, train teachers, and revise curricula continuously. Experience at the high-school level in designing and teaching interdisciplinary courses is sparse. New inservice programs and support will be required, as will modification and improvement of the curriculum.

Recommendations

- **Materials and syllabi for the capstone course need to be developed and tested before widespread adoption can be expected. Curricula for a variety of topics should be developed and tested, with models for inservice support for teachers. New materials should be developed, and existing materials identified and modified. Carefully designed evaluation problems to** assess student outcomes should be part of the development and field-testing program. With appropriate foundation or other support, the development of the course could occur through a program of competitive grants to high schools, local school districts, or partnerships between the latter and interested university faculty or industry scientists. Some models for such a course exist in colleges and universities, and that experience should be exploited wherever possible. Where available, regional mathematics-science centers could participate in the design, testing, and evaluation of pilot programs, as well as revision of actual programs. Sufficient financial support should be provided to ensure not only the introduction of projects, but also their long-term monitoring, evaluation, revision, and the necessary inservice opportunities for engaging additional teachers.

• Accompanying the development of modules for the capstone course, there needs to be an overarching process of evaluation that not only identifies the best modules, but provides a mechanism for their widest use. That means not only making materials available, but providing guidance and support for teachers who are new to the program. The involvement of more than one teacher and the use of resource people are highly desirable and should be the case wherever possible. Few teachers, even at the university level, are comfortable in taking on such a course by themselves, and part of the message that should be conveyed to students is that people must cooperate in addressing complex issues.

THE ROLES OF SPECIAL SCIENCE SCHOOLS AND CENTERS

Several types of specialized settings offer intensive programs in science and mathematics, usually for talented and "gifted" students. At the high-school level, they can be loosely categorized as follows:

• Traditional urban public high schools that offer specialized curricula in science and mathematics.
• Newer urban public schools referred to as magnet schools.
• State-sponsored residential schools of science and mathematics.
• Local and regional centers for science and technology that present science courses. Students attend the centers for part of the school day and participate there in a wide variety of activities involving science and mathematics.

Older, Specialized Public High Schools

Some large urban areas have long publicly supported high schools with college-preparatory curricula that emphasize science and mathematics. Admission to those public high schools is highly competitive and is often based on results of entrance examinations or other measures of performance or ability. The schools tend to serve "gifted" students. They offer more laboratory work than regular schools, and many are linked to local firms and research laboratories that provide equipment, mentors, and opportunities for participation in research (OTA, 1988). Teachers are encouraged to devise new curricula and to develop new teaching materials in collaboration with their colleagues. New York City has several long-standing examples: Bronx High School of Science, Brooklyn Technical High School, and Stuyvesant High School. Philadelphia's Central High School, although not exclusively for science and mathematics, provides talented students with a wide variety of enriching activities, opportunities for independent study, seminars, and extracurricular experiences. Baltimore Polytechnic Institute is another well-known example.

Magnet Schools

Magnet schools were recently introduced as vehicles for desegregation, as well as improved education, and are playing an increasing role in urban systems. They provide opportunities for all students to enroll in programs that interest them, rather than restricting entrance on the basis of ability. Most magnet schools share several characteristics. First, they feature a special curricular theme or method of instruction, which in some instances focuses on science and mathematics. Second, within a district magnet schools play a role in voluntary desegregation. Students and parents can choose a school, and there is open access to students from beyond the immediate school zone (Blank, 1989). Magnet schools can be found at the elementary, middle, and secondary levels.

Magnet schools have grown markedly in number and influence, particularly in the last 5 years, and there are now more than 1,000 (OTA, 1988). According to Blank (1989), the average urban school district with a magnet-school program has over 50% more students in magnet schools than in 1983. In the average urban district, about 20% of students are in magnet schools, and the demand is increasing.

A national study identified four major factors contributing to the growth of local interest in magnet schools (Blank, 1989, p. 4):

• Development of a voluntary approach to school desegregation.

• Interest in educational options and diversity in curricular offerings (such as advanced programs, arts, science, and foreign languages) and in school organization (such as alternative schools, open schools, traditional or basic education, and individualized instruction) with the objective of improving the overall quality of education in a district.

• Greater attention to the outcome of public education, including preparation of students for careers and preparation for decisions on further education or training.

• Renewed concern with the quality of education on the part of community leaders, parents, and educators, as exemplified by the well-known report of the National Commission on Excellence in Education, *A Nation at Risk* (1983).

Magnet schools advance educational equity by attracting students with common educational interests, but diverse abilities and socioeconomic backgrounds. The heterogeneity of students is accomplished by providing educational experiences generally not available in the other public schools in the area.

Studies have been conducted to determine whether the goals of magnet schools have been reached. In some cases, assessment involves only the effects of magnet schools on desegregation and on the choice and diversity of curricular offerings; such assessment shows that magnet schools meet these goals. Few districts, however, assess the deeper educational effects of magnet schools, and most are content if the schools meet the mechanical objectives of the program. For example, interest in assessing the educational effects of magnet schools on students with a wide range of backgrounds and abilities is usually modest, if the district's primary motivation for magnet schools is desegregation (Blank, 1989).

With the growth of magnet schools as an important element in urban public education, several important policy issues have emerged. Although magnet schools were designed to serve students who seek the opportunities offered by choice and diversity, there is growing concern that magnet schools do not serve students who are at risk or students who are most likely to drop out of school. Thus, the goals of educational equity are not being met. Some are also concerned that the popularity of magnet schools is causing a division of public education into two tiers: a set of schools that offer special opportunities for some students and neighborhood schools that offer education of lower quality for the remaining students (Blank, 1989). Those issues are sharpened by the lack of hard information on the educational accomplishments of magnet schools. Research to address the latter question requires sophisticated analysis of many variables—measures of student outcomes in both magnet and nonmagnet schools, longitudinal studies of both student populations, analyses of district and school policies and organization, and so forth (Blank, 1989). Few districts have conducted such a study, but, as costs of magnet schools increase, studies will be needed to justify increased expenditures.

Residential Schools for Science and Mathematics

A relatively new experiment in fostering quality education is the residential school for science and mathematics. Six are operating, and plans are being made to open residential schools soon in several other states. Admission is highly selective and is based on results of admission tests and high ability in the sciences and mathematics. Students are drawn from schools throughout their own states. The schools are state-supported, and instructors, also chosen from a highly competitive applicant pool, are given free reign to develop the curriculum. With one exception, the residential schools offer 2-year intensive programs in science and mathematics that are supplemented by core courses in the humanities; the school in Illinois is a 3-year institution. In addition, several of the schools plan to serve as resource centers for inservice training of teachers and as centers for developing and testing new science curricula and laboratories. To facilitate exchange of information among the schools—what is working and what is not, sources of additional funding, and ideas for improving curricula—they have formed a National Consortium for Specialized Secondary Schools of Mathematics, Science and Technology, headquartered at the Illinois Mathematics and Science Academy.

The residential schools have been in existence only for a short time; the oldest, the North Carolina School for Science and Mathematics, was established in 1980. Therefore, there are few data indicating whether students who attend them go on to study science, mathematics, or engineering in college. A recent survey, however, found that 80% of the graduates of the North Carolina School majored in science and engineering in college (OTA, 1988).

Centers for Science and Technology

Centers for science and technology offer another alternative for students interested in science. They are private and publicly funded regional centers that offer advanced courses in science and technology to students from many high schools. Students usually spend half their in-class time at their home schools and attend the science and technology centers specifically for their science classes. In some instances, they can earn college credit. Students are also encouraged to participate in scientific research projects with local mentors. Some centers have begun to develop new curricula and instructional materials and serve as resource centers for local high-school teachers.

Appendix E lists examples of each type of school discussed above.

Recommendations

• **The relative autonomy of both state-sponsored residential schools for science and mathematics and the centers for science and technology provides a unique opportunity for these institutions to serve as "laboratories" for curricular reform. In addition to providing high-quality instruction, they should be encouraged to continue in the development of new curricula, instructional materials, and techniques for assessment. They can also serve as inservice centers for local high-school teachers. For educational experiments to have maximal impact nationally, mechanisms should be devised for comparing and assessing the programs at the residential schools and regional centers and disseminating the resulting information broadly.**

• **Research is required to assess the educational effects of magnet schools both on their students and on the associated neighborhood schools.**

8

Achieving National Goals:
Dilemma and Resolution

WHAT HAVE WE LEARNED FROM THREE DECADES OF ATTEMPTED EDUCATIONAL REFORM?

By one estimate, more than 100 committees have met, promulgated solutions, and disbanded since the publication in 1983 of *A Nation at Risk*—and the failings of science education have not diminished to any measurable degree (Hurd, 1989b). The specific suggestions tend to have a familiar ring, yet nothing very much changes. Twenty years before, there had been some substantial efforts to alter the teaching of science. Prompted by the USSR's launching of Sputnik in 1957, the nation mounted massive and expensive efforts to improve science curricula. Although some of the results were impressive, they were always ephemeral. If reform is to be successful now, we must understand why earlier attempts have been so ineffective.

The major project for the reform of biology education emerged in the late 1950s as the Biological Sciences Curriculum Study (BSCS), which was initially part of the American Institute of Biological Sciences. At the outset, it sought to be concerned with biology education in grades K-12; a parallel project, the Commission on Undergraduate Education in the Biological Sciences, was to deal with biology in the colleges and universities. That holistic approach was abandoned, however, when the funding agency, the National Science Foundation (NSF), restricted the BSCS's initial efforts to the tenth-grade high-school biology course.

In a surprisingly short time, the BSCS produced and began to test three versions of a high-school biology curriculum. After 2 years of testing and revision, the three versions were produced by commercial publishers, and they soon captured a major portion of the textbook market. Although each version was different, the approach was the same: reduce memorization, make inquiry the mode of teaching and learning, concentrate on themes and concepts, emphasize hands-on laboratory and field work, and relate biology to human problems.

The BSCS versions soon dominated biology education and were imitated by many other textbooks. It was generally agreed that a major improvement in biology education had emerged. But the movement was not sustained. In this country, the BSCS programs slowly decreased in number, and today they are used in only a small percentage of the schools.

Why did this effort to reform biological education dissipate? A principal reason is that too few teachers are both skilled in inquiry-based teaching and broadly knowledgeable in their discipline. A major failure of the BSCS was its inability to continue to prepare teachers to use the BSCS programs effectively. The original BSCS textbooks were far more demanding of both teachers and students than were the prevailing textbooks, and essentially all teachers who joined the BSCS program needed additional education. The BSCS was able to give brief refresher courses only to teachers who were involved in testing the experimental editions. NSF did not support the BSCS in offering inservice summer institutes and refresher courses on a large scale—at least in part because of political sensitivities aroused by rival publishers—although for educational reasons such offerings were required. Some individual biologists did offer BSCS-like refresher courses, but they were not directed by the BSCS.

The failure of adequate inservice support might have been overcome in time if the institutions training new teachers had taken the responsibility for teaching how the BSCS materials could be used effectively. But almost no institutions responsible for the education of teachers paid attention to the BSCS or to any other national efforts to reform science curricula (Mayer, 1986).

There are additional reasons for the decline of influence of the BSCS. Despite the intention of the original BSCS team, the program really addressed college-bound students and failed to deliver "science for all." Moreover, the effort at reform failed to take into account many of the obstacles we have discussed earlier in this report: the complexities of textbook adoption, the ambiguous role of NSF and other federal agencies in an educational system based on local control, and the roles of testing, college admissions, and school administrators (Jackson, 1983).

A second major effort was the NSF-sponsored summer institutes, which involved large numbers of concerned university scientists working with gifted high-school teachers and administrators and costing hundreds of millions of dollars in public funds. As noted earlier, many participants felt that they were both effective and important, but others disagreed, and for a variety of reasons the institutes were abandoned in the late 1970s. A subsequent review (GAO, 1984) has claimed that they did not have a measurable effect on student performance—a result disputed by many others who state that this concern was based on superficial analysis. Although there might be argument about their effect on student outcomes, the institutes had a positive and lasting effect on the morale and sense of professionalism of the teachers who participated.

THE NEED FOR NATIONAL LEADERSHIP
AND FEDERAL FUNDING

The picture we have painted throughout this report is daunting and depressing in its complex relationships. Textbooks will not improve unless market forces change. Market forces will not change unless professional educators and the public accept new formulas for measuring the success of our school system. New measures of success must be accompanied by different goals of teaching and learning, which require not only different kinds of teaching materials, but different styles of teaching. More imaginative and effective approaches to the classroom require that teachers themselves learn differently in colleges and universities, and that can happen only if the scientific community sees itself both as part of the problem and as part of the solution. Changing the perspectives of teachers now in service requires massive efforts of everyone: teachers, scientists, test-makers, schools of education, publishers, school boards, teachers unions, and parents. Some of the major relationships are depicted in Figure 1.

At every point and on every issue discussed in this report, obstacles present themselves. Someone will be inconvenienced, challenged, asked to sacrifice, and forced to rethink what is being done and how. Furthermore, the incentives for improvement, if any, are minimal, if only one or a few parties are willing to change.

The kinds of changes called for in this report will occur only if major federal funding is committed to make them happen. It is unreasonable to expect local school districts, with their many competing priorities and lack of scientific expertise, to finance all the actions required to effect the changes needed to turn the country around on the important issues discussed in this report. The problems we have been discussing are national ones that seriously affect the future of the United States. Only a concerted and highly visible national program can hope to mobilize the many different constituencies that must participate. Major changes in direction will occur only if the states and local districts are engaged through the availability of federal funds that are specifically targeted to causing revolutionary improvements in K-12 science education.

What level of funding will be required? The committee developed a number of different estimates of the magnitude of resources required (see Appendix G). Although it did not analyze in depth the costs of its recommendations, it is clear that any effective change will be expensive. The call for a much greater financial commitment than currently exists is coupled to the recognition that leadership in science education must be exerted by the scientific community to ensure that the new resources are effectively spent. Moreover, successful change will require sustained, coordinated actions by many groups, as well as continuous monitoring and evaluation to guide the process.

It is important to note that the previous efforts to reform science education were also expensive. For example, the NSF-sponsored summer institutes cost hundreds of millions of dollars over two decades (GAO, 1984; Hooper, 1990). At least this level of inservice activities is needed in the near term. Additional funds will also be necessary to address simultaneously the other parts of the

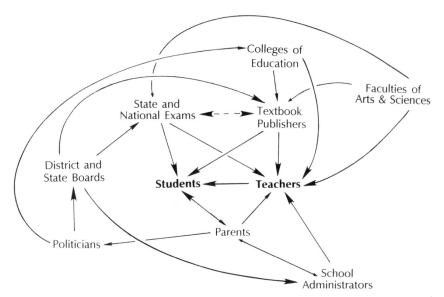

FIGURE 1 Directions of strong and moderate influences at several nodes in the web of public education. At the center of the network are the students, for whom the web exists, and the teachers, on whom success depends. Note, however, that the influences on both are largely beyond their control. When the influences on teachers are not supportive, this set of relationships tends to reduce the role of teachers from knight to pawn. The arrangement also makes significant educational change very difficult, as several other nodes in the network, each far removed from direct involvement in the classroom, must be perturbed simultaneously. The response of each of these nodes is in turn determined by this distinct set of influences impinging on it.

The position of the faculties of liberal-arts colleges and universities is of particular interest. Well placed to see the results of educational failure in the schools, thus among the first to criticize when failure ensues, and enjoying the greatest autonomy for corrective action, these faculties have been among the last to recognize in any organized way the role they must play if change is to be effective. Their influence on high-school teachers is strong (because at the outset they teach the science to teachers), but this impact is neither as useful as it ought to be nor sustained through subsequent contact, such as inservice activities. The influence of scientists on high-school texts and tests is so feeble and indirect as to be virtually nonexistent, and support for colleagues in schools of education is usually disdained.

From the styles of each, the influence of texts on tests and vice versa appears to be strong and direct, but both may be responding to the same set of outside forces. In effect, however, texts and tests have been mutually reinforcing. In a hopeful development, national testing agencies have recently indicated plans to break from tradition and develop new forms of examinations that will be better designed to assess the ability of students to reason (National Governors' Association, 1990).

The diagram does not include all the important players; for example, teachers unions and foundations interested in education are not shown. Furthermore, the strengths of interaction may differ locally; for example, some schools have involved parents to a greater extent than suggested here. And, finally, some important interactions are not depicted; for example, curricula are often influenced by college admission requirements, and the behavior of teachers can be strongly influenced, for better or worse, by the perceptions of colleagues.

complex system—testing, preservice education, textbooks, other instructional materials, laboratories, and research.

The magnitude of the task should be seen in perspective. The nation spends an estimated $350 billion on education each year (U.S. Department of Education, 1989), significantly less than the reported $500 billion for industrial capital investment (*New York Times*, April 11, 1990). The estimated $94 million per year (in average 1989 dollars) for inservice programs for biology teachers (Appendix G) is only 0.027% of the nation's education budget. Given the magnitude of the education enterprise, this figure represents a very modest commitment to meaningful change. Similarly, the government spent about $124 million in 1987 on educational research (GAO, 1988). Considering the central importance of education to our future economic prospects as a nation and the consensus that our educational system is in trouble, this is a very meager investment. No industry would expect to compete, let alone retool, with such a token investment in the future; we need the equivalent of extensive retooling if we are to succeed in changing how our children learn.

A ROLE FOR THE SCIENTIFIC COMMUNITY THROUGH THE NATIONAL ACADEMY OF SCIENCES AND THE NATIONAL RESEARCH COUNCIL

How can our massive decentralized educational system be perturbed in ways that will lead to more effective teaching and learning of science? All the needs for change enumerated in this report have been identified in previous studies; the trumpet of reform has been sounded many times in the recent past. The lesson of history is that nothing short of a concerted effort on all fronts, sustained by visible and effective leadership, will be successful. What can be done to encourage, if not ensure, a broad attack?

The presence of an overarching body capable of focusing the attention and galvanizing the interest of all concerned on science education seems essential. To play a leadership role, such a body should fit several criteria:

• It should be composed of scientists, science educators, and teachers and should be capable of mobilizing expertise of the highest quality in both science and education.

• It must have sufficient prestige for its studies, analyses, and recommendations to command wide respect on a national scale.

• It should not be associated exclusively with any particular group of vested interests, such as schools of education, teachers unions, or scientific societies with focused concerns, yet it should be able to invite their participation.

• It should remain independent of the federal government, immune to political pressures, and therefore able to assess, diagnose, and recommend freely.

We are skeptical that such a body could be created de novo, for it would lack the necessary intellectual authority and prestige. Its voice might be heard, but, like the voices of so many others, to little effect. Few existing institutions meet the criteria. Perhaps the one that comes closest is the National Academy of Sciences (NAS). Under the aegis of NAS, the National Research Council (NRC) already has a board devoted to the state of education in mathematics. The growing recognition and concern within the scientific community, the education community, and the general public about science education make it critical that the scientific community act now to improve science education in the United States. *It would be both precedented and appropriate for the National Academy of Sciences to create in the National Research Council a Commission or Board on Science Education to monitor and to provide advice on the status of teaching and learning in science.*

How could such a body help? First, through continuity of attention. As we have described, after the USSR's launching of Sputnik, the United States undertook to reform the teaching of science in the schools, but little remains of that effort. In our country, political enthusiasms have short lives. No organization today has a continuing involvement in the analysis and evaluation of science education nationally, serves as a catalyst for dialogue among the various constituencies concerned with science education, and can provide appropriate advice to Congress, the administration, the states, universities, and the general public on science education. The changes that must be made in our educational system will take years to effect, and they will not occur unless they are monitored and encouraged throughout the process. NRC can play a role in those activities.

Second, research scientists must become more engaged on a continuing basis in actively fostering ways of improving the preparation of teachers in the sciences, in helping to identify needs for research in science education, and in providing criteria and guidelines for the evaluation of texts and other instructional materials. By virtue of its stature in the scientific community, NAS, through NRC, can effectively call on scientists to become actively involved in collaboration with teachers and science educators.

Third, the scientific community, through NRC, can provide advice about the distribution of funds to the various areas that need to be addressed to improve science education. Through a continuing collaboration among scientists, teachers, and science educators, such a board would constitute a unique forum for identifying the areas needing the most immediate intervention and for recommending and evaluating activities responsive to those needs.

Creating a board that will be able to operate effectively will not be a simple task. Unlike the mathematics community, the scientific community is fragmented into many disciplines that rarely discuss with each other questions of either instruction or curriculum. Furthermore, as we have pointed out in several places in the report, the ties between the primary and secondary education communities and the science faculties in colleges and universities are weak and in need of repair. Reaching agreement on what issues should be addressed in the classroom, how these issues should be approached, and how we are to measure progress will require sustained effort and cooperation. Overcoming these manifold difficulties, however, is central to the challenge confronting science education in the United States. While stressing the urgent need for action, we do not underestimate the obstacles. But we see a board within NRC offering an especially promising opportunity for building bridges between the science and education communities—bridges that will facilitate the kind of broad consensus that is essential if we are to achieve quality education in science.

Throughout this report, we have pointed to instances of questionable educational practices that contribute to poor learning. Let us reconsider some of them here, exploring how a board of the National Research Council—or some equivalent body—could contribute to effective change in how science is learned.

• There are no common standards or established criteria based on educational research to guide the creation of new *inservice programs*. But, as we saw in an earlier section, extensive inservice experience is essential in the near future for most teachers now in the classroom. There are clear needs for the development and testing of pilot programs, for analysis and evaluation of the effectiveness of current programs, and for the wide dissemination of information about successful models. Those are all activities that NRC, with the support of foundations and industry, can help to promote.

• Another role of a new body would be the creation of criteria for evaluating the effectiveness of *preservice programs* for teacher education. The current preservice instruction of teachers does not convey knowledge of science in a manner that is helpful to future teachers of children. The mode of instruction is devoid of "content pedagogy"; it does not assist teachers in relating scientific concepts to the knowledge that a child brings to the subject. On the other side of the instructional coin, preservice experiences in pedagogy are largely divorced from the content of the science. NRC can explore the barriers that now prevent scientists and science educators from working together in training student teachers.

• There are no generally agreed-on standards for *textbooks, state and national tests,* and *curricular content.* Some texts, tests, and curricula might be good, but there is little third-party effort to examine their appropriateness and success in meeting basic goals in science education. We are not suggesting that a body like that proposed prescribe the content of textbooks. Quite the contrary. In a truly excellent educational system, teachers would have great freedom to use the books and techniques that suited them. At a minimum, however, such a body might develop criteria and standards for the quality of textbooks and even provide a periodical "consumer's guide" to the books on the market. These books would be examined thoroughly and critically for accuracy, readability, coherence, and effectiveness in conveying science as a process and as a way of knowing. It could also give guidance to persons responsible for selecting textbooks for states and school districts. Such information should consist not of individually written and potentially idiosyncratic book reviews, but of broadly based input from both teachers and scientists, presented against a backdrop of educational goals that are themselves subject to review and discussion. We see such an effort as valuable to teachers and publishers alike. If it helps to steer the market and thereby the production of higher-quality books, it will improve the education of students.

The possible agenda of an NRC board or commission with responsibility for monitoring the health of science education nationally is so large as to be overwhelming; but that speaks to the importance of the task. Other potential functions and projects include the following:

- Evaluation of the role of national and state *examinations.*
- Analysis of models for *integrated science curricula* that start in the early grades and build in a coherent way through high school and into college.
- Promotion of interdisciplinary cooperation in the *preparation of teachers.*
- Creation and enhancement of mechanisms for the *collection and dissemination of information* on all aspects of science education, perhaps including computer-based networks or even a series of regional centers that teachers could visit to acquire hands-on experience with new materials and laboratories.
- Evaluation of *advanced-placement curricula and examinations.*
- Identification of *research needs in science education.*
- Identification of new ways to promote *professionalism* in the community of teachers.
- Identification of new ways to *interest students from various ethnic backgrounds in science and teaching as career options* and assessment of *more effective ways to teach them.*
- Stimulation of wider appreciation for the *role of science in society.*

The present lack of leadership and unity of purpose in science education is illustrated by the fact that the important explorations of educational reform now in operation are not yet well coordinated. Perhaps the most ambitious is Project 2061 of the American Association for the Advancement of Science (AAAS, 1989), which seeks to reform the teaching of all the sciences and technology throughout grades K-12. It has formulated general recommendations and is now designing courses. The AAAS Project on Liberal Education and the Sciences is targeting the college and university courses intended for nonmajors. A group at Stanford University is developing a 2-year biology sequence for middle schools based on its undergraduate human-biology program. The Science as a Way of Knowing project, sponsored by the American Society of Zoologists and 10 other societies and organizations, is producing materials for teachers of college introductory biology courses. The Scope, Sequence, and Coordination project, recently developed by the National Science Teachers Association, is an attempt to offer more science in grades 7-12 (Aldridge, 1989). And many other organizations, such as the BSCS and the Education Development Center, continue to develop new curricular materials for various levels.

In summary, the reform of science education will require national leadership from the scientific community and a major investment at the federal level. National leadership is needed to develop a national consensus that will press for the production of well-educated teachers, to insist on fine textbooks and supplies, to identify outstanding model curricula and inservice programs and make them available to all schools, to mobilize the universities so that support for precollege science teachers becomes one of their major goals, to encourage the professionalization of teachers, and to encourage cooperation of parents and other adults in the education process. National leadership must offer a vision of science education that enables our system to produce students prepared to face the challenges of the twenty-first century.

Executive Summary

Management is the capacity to handle multiple problems, neutralize various constituencies, motivate personnel; in [schools], it means hitting as well the actual budget at break-even. Leadership, on the other hand, is an essentially moral act, not—as in most management—an essentially protective act. It is the assertion of a vision, not simply the exercise of a style: the moral courage to assert a vision of the institution in the future and the intellectual energy to persuade the community or the culture of the wisdom and validity of the vision. It is to make the vision practicable, and compelling.

A. Bartlett Giamatti (1988)

Educational leadership, as Giamatti understood, is difficult business. Difficult enough in a university, the task is monumental in a larger pluralistic society where so much of education is under local influence and control. By whose authority does one lead? And how is the band inspired to play, let alone follow? As Giamatti asserted, vision is the key; great obstacles can be overcome if there is a coherence and logic and purpose to the vision.

This report offers a vision of what our schools might become: places in which all the nation's children are taught about science from the earliest years in such a way as to awaken curiosity and wonder and appreciation for the world in which they find themselves. And it is about great obstacles—greater obstacles than most advocates of educational reform recognize. But it is also about purpose and the need to see the difference between leadership and management. A major theme of this report is that much of what passes for educational reform is not leadership; it is tinkering with management.

This is an optimistic report. At some moments in history, societies are susceptible to important change, the crescendo of inadequate performance begins to sound like failure, and the players are receptive to new ideas. In various respects, particularly in science education, the nation's system of schooling has crossed that threshold.

This report is explicit about a number of problems. Many (perhaps all) have been recognized by previous committees and panels, but not much seems to have changed. The centerpiece of this document therefore is the need for leadership, specifically the need for leadership from the scientific community. It is not that scientists and scientific organizations have failed to contribute to the clamor for change. Indeed, on occasion there have been significant and effective contributions to primary-school and secondary-school programs from scientists in universities, industry, and government. The problem is that, like virtually every other effort at reform, they remain local and isolated contributions, unguided by any overarching plan, unaccompanied by any independent assessment, untouched by any means of propagation, and, hence, ephemeral. We seek a program for sustained reform.

Our most important recommendation is directed to the scientific leadership of the nation. This committee has developed a strong sense of what must be done to improve science education in the years from kindergarten through high school, and this document lays out both problems and means of solution. We are certain, however, that the necessary changes cannot be made unless there is a permanent organization to monitor and organize them. Moreover, the leadership provided by the organization should be separate from its multiple constituencies. We feel it necessary that the scientific community be involved as both guide and goad, both resource and participant. Specifically, we urge the National Academy of Sciences to create in the National Research Council a permanent board or commission with the charge of monitoring and improving the state of science education in the United States. We do not propose, however, that the scientists undertake this task in isolation. The organization we envision must be a cooperative venture that fully involves knowledgeable science educators and outstanding teachers at all levels.

The challenge that faces us is enormous, and effective change will require participation of individuals from national to local levels. The establishment of a permanent Board or Commission on Science Education is therefore only a start in the right direction. The specific tasks to be assumed by the new body and the necessary initiatives that must be taken by other institutions concerned with education constitute a major part of our report. We have also described what understanding of biology will be required by Americans in the coming years and what institutional practices must change at all levels to ensure that this understanding is achieved. We examine both the obstacles and the potential resources available to overcome them. In a number of instances, our definition of specific problems is incomplete, because there is a compelling need for research and analysis as well as guidance and leadership.

Before setting forth an agenda for such leadership, let us summarize the present condition of science education in our schools and what needs to be done to see improvement. At the outset, we acknowledge that conditions in a number of schools, particularly in urban areas, make it nearly impossible for teachers to teach and for children to learn. Where those conditions are present, society's first priority should be to correct them, and we do not underestimate the difficulty of the task. But this report looks beyond that first line of attack and addresses issues specifically related to the teaching and learning of science in general and biology in particular.

Science—some knowledge of nature, some understanding of how knowledge about the world is obtained, some feeling for the relation between cause and effect, for uncertainty, for the magnitudes of space and time—should be presented in the earliest years of elementary school, but it is not. We do not have a teaching corps trained for the task, and we introduce students to science in the middle schools or junior high schools and continue in the high schools with a curriculum that seems almost designed to snuff out interest. Inadequate science education has important economic and political implications for our nation, for citizens must be equipped with the ability to make informed judgments about health and social issues, and the demands of an increasingly technological economy will require a better-educated work force.

We do not start teaching science early enough, and we reduce science to a language of foreign terms to be memorized. We fail in the middle schools to recognize the interests and stages of development of the young adolescent. Those mistakes deprive students of the opportunity to use scientific concepts and related facts to respond to issues of health, such as alcohol and drug abuse, and broader issues, such as environmental pollution, and they discourage able students from entering scientific and engineering professions.

We reinforce the students' sterile experience with standardized tests that require the recognition of terms, and we teach from textbooks that are often too long, are abysmally crafted, and are written with little or no understanding of biological concepts and with the implicit aim of avoiding offense to a scientifically illiterate segment of the adult population.

Too many of our teachers are recruited from the lowest academic quartile of the college population, and in the liberal-arts colleges and universities prospective teachers are themselves taught in lecture formats that ill prepare them for their future role. Schools of education present pedagogical information so unrelated to the specifics of teaching science as to be of little or no help. The opportunities for teachers of science to update their scientific knowledge and skills and to interact with each other and with research scientists—once a prominent part of inservice training—have withered, victims of misguided policy at the federal level. And the teaching of science in an adequate laboratory environment is available to few students.

As a result of those interlocking practices, science usually is presented poorly and inadequately learned. Middle-school and high-school students do not learn that science is a process of inquiry about the world, and they do not become engaged themselves in developing an understanding of scientific concepts. They memorize; because what they have memorized seems irrelevant to their world, they soon forget. Worst of all, most students leave the experience with the conviction that further exposure to science is something to be avoided if at all possible.

We offer a large number of specific recommendations to address these and related matters. They are summarized here, grouped in 12 categories in the sequence in which the background and reasoning appear in the body of the report. The final category, "Leadership," returns to the need for direct participation of the scientific community, in formulating goals and in creating mechanisms to measure progress toward achieving these goals.

The Curriculum

The last 20 years have transformed the United States into a society that is increasingly dependent on science and technology, but the transformation has barely permeated our system of education. For deeper understanding to occur, science must be treated as a high-priority subject. Like mathematics, reading, and writing, science should be a core subject whose process must be comprehended by all.

- Beginning in the formative years of elementary school, substantially more time needs to be devoted to science. The biological science presented to young children should have natural history as a major focus, be integrated with other subjects wherever possible, and emphasize observation, interpretation, and hands-on involvement, rather than memorization of facts. To prevent the acquisition of detailed factual knowledge for its own sake, achievement tests, when used, should stress conceptual understanding and development of skills. Reading and writing about natural phenomena, appropriate for the range of readers in elementary-school classrooms, should be an integral part of language-arts and reading instruction. Many states have already developed science frameworks or curricular guides outlining the amount and type of science that should be taught at all levels, but states, districts, and especially schools must ensure that the requirements are instituted. This will require much more than token observance of new regulations.
- Because children of middle-school age are curious about their bodies and full of misconceptions about health, hygiene, and disease, an orientation to human biology holds great promise both for sustaining students' interest in science and for addressing a variety of educational goals important to society at large. Several groups are developing such courses, and their results should be compared and evaluated by the science and science-education communities.
- At the high-school level, the central concepts and principles that every high-school student should know must be identified, and the curriculum pared of everything that does not explicate and illuminate the relatively few concepts. Those concepts must be presented in such a manner that they are related to the world that students understand in language that is familiar, and they must be taught by a process that engages all the students in examining why they believe what they believe. This requires building slowly, with ample time for discussion with peers and with the teacher. Particularly in science it also requires observation and experimentation, not as an exercise in following recipes, but to confront the essence of the material.

Textbooks

Most available texts are poor, but the problem of biology textbooks intersects with other issues discussed in this report. We look to teachers to provide instruction in science for our young people, and in this our teachers need enormous help and support. Improving the textbooks requires a greater emphasis on the place of concept and process in teaching biology, as well as a clearer picture of the goals of science education from kindergarten through high school. Achieving that emphasis needs consensus among teachers, changes in how student accomplishment is measured, and changes in the expectations for texts that state boards convey to publishers. If those changes can be accomplished, the publishers will find it in their interest to produce better texts.

• Extensive review of science texts should be instituted by the scientific community. Textbooks need to be assessed for scientific accuracy, currency, and vision, and the reviews will need to be widely available to teachers, members of school boards, and others at the grass-roots level. The broader scientific community should collaborate with teachers in evaluating textbooks and, on a local level, provide advice on textbook adoption. It is important that evaluations include input from scientists, people experienced in the school classroom, and researchers in learning and reading comprehension.

• Scientists should be engaged in the writing of middle- and high-school texts, and control of content shifted from publisher to author. It is essential that sufficient time be devoted to the project so that adequate classroom testing can be done and analyzed before books go to press. Particular care should be given to designing smaller texts around important biological concepts and principles. Technical language should be used sparingly and never as a substitute for lucid explanations of biological processes. Illustrations not only should be accurate, but should be designed to increase understanding.

Laboratory Activities

Properly designed laboratory activities are essential for effective biology courses. However, activities that merely illustrate what a text has presented do not produce the desired results—promoting interest, curiosity, and understanding. The prevalent form of laboratory activity must be replaced by genuine investigations, designed and tested to enable students to achieve the conceptual changes necessary for intellectual development and understanding. Laboratory work and field work are central to a major reconstruction of biology education.

• A major effort should be initiated to identify current exemplary laboratory activities for the biology curriculum. Laboratory activities should take advantage of recent research about how students learn science and should contribute to the development of fundamental biological concepts. The effort to find such activities could be underwritten with public and private funds and carried out in conjunction with model inservice programs.

• Groups should be assembled to develop and assess model laboratory activities. Such groups should include biology teachers, university research biologists, and researchers in science education. The inclusion of students would be essential to test the activities. The groups should not only design and test laboratory activities, but also develop appropriate measures and indicators of the effects of laboratory work and field work on student understanding of biology. The groups' work could take the form of summer workshops at research universities.

• A system should be developed for providing inservice education in the use of laboratories. It could take the form of regional meetings for teachers in the summer or be conducted in cooperation with school-district inservice programs. Such teacher education would enable teachers to be laboratory students and to work through the laboratory activities with people who can interweave laboratory experience with effective teaching, providing a model as to how a particular activity is approached most effectively.

• Laboratory activities require more time than normally allotted. They will not be able to occupy their appropriate place in the curriculum until time is created to accommodate them.

Tests and Testing

Understanding of central concepts and principles of biology will not be attainable as long as current classroom tests and standardized tests assess only recall and recognition. Tests that are consistent with a new commitment to understanding principles and concepts are essential to enable teachers to know what they are accomplishing as they change their teaching methods and emphases. They are also necessary to inform students that different learning strategies are needed to achieve the goals of their biology course. Testing is increasingly driving curriculum and instruction in a dull and pedantic fashion, so it is imperative to address the issue of testing and evaluation in biology at all levels—national, state, school-district, and classroom. If appropriately developed, tests might well continue to drive the curriculum, but in ways that are in the best interests of students.

• A new array of test instruments and procedures should be made available to enable biology teachers to evaluate and improve their teaching and their students' learning. Methods need to be designed specifically to address how well schools are doing and be sensitive enough to show how students' performance differs when teaching changes.
• The nation should be concerned more with evaluating the effects of curricula, teaching methods, and materials than with ranking the performance of individual students. The system's components should be probed, rather than just the relative ranks of the learners.

Other Factors That Hinder Effective Education

Substantial improvement in the teaching of science will require change in school administration. More flexibility is required in the scheduling of classroom and preparation time in the pursuit of related professional activities by teachers and in teachers' sharing of responsibilities. Fiscal implications are inevitable. True educational reform will rock many boats, and those who must pay for change should be clear about the goals they wish to achieve.

- Obstacles to creative teaching must be lifted. Inasmuch as textbooks and testing play important roles in determining how biology is taught, teachers must be encouraged to experiment with new techniques in pedagogy and assessment. School policies, rather than perpetuating isolation, should be tailored to encourage teachers to work together in developing ideas. Opportunities for teachers to visit other classes, participate in inservice programs, or teach cooperatively must be made available by released time and the use of mentor teachers.
- Teachers should be enabled to devote nonteaching time to activities that will enhance their ability to convey knowledge to their students, such as preparing laboratories and tutoring students.

Preservice Education: Teaching the Teachers

The preparation of teachers needs drastic reform. Current standards for content and pedagogy are inadequate to meet society's expectations. The situation will worsen in coming years, unless teacher preparation becomes much stronger. Effective biology teaching requires being able to do, as well as to know, and new programs must ensure that teachers not only understand biology, but have the skills to relate scientific concepts to children of different ages.

- A curriculum that treats science as a process for knowing about the world can be effective only if the teachers have a deep understanding of that process themselves. We therefore feel that every teacher who has responsibility for a high-school science class should have had the experience of engaging in original research under the direction of a research scientist. Ideally, this happens as part of preservice education, even if only for a semester or a summer. For practicing teachers who have missed the opportunity, inservice mechanisms must be devised.
- Prospective teachers of high-school biology need adequate preparation in cell and developmental biology, ecology, evolution, genetics, and molecular biology and biochemistry. Those fields should guide their selection of courses and should be supplemented by exposure to mathematics and the physical sciences. We encourage experiences that explore new ways to break down the traditional barriers among the natural sciences and between the natural and social sciences.
- University science departments and schools of education bear major responsibility for problems related to teaching of teachers. Neither has provided course work appropriate to teaching science at the K-12 level, and liberal-arts faculties have not encouraged their best students to consider teaching as a career. The most important change in the undergraduate curriculum will be to require the participation of university faculty in creating environments for learning that are less authoritarian and that engage future teachers in discussions of concepts,

study of the relations between scientific disciplines, and cooperative analysis of information.

• New processes should be developed for integrating pedagogical and scientific subject matter more effectively. Schools that train many teachers could create sections in which students have an opportunity to discuss how their experiences at the university level could be best used as a foundation for presenting important concepts and principles to younger age groups. Collaborations should be developed between faculties of schools of education and science departments to develop science-methods courses and to improve pedagogy in undergraduate natural-science courses. The goal of such courses would be to combine appropriate teaching methods with scientific method, and they would be taught by scientists or science specialists.

• Undergraduate programs are needed that will better prepare teachers to deal with science in elementary and middle schools. Such programs could have an integrated science or science-mathematics major. The pedagogical character of the programs will differ from that appropriate for high-school teachers, but there are few if any usable models.

• Research is needed on what makes education programs for teachers effective.

Licensing and Certification of Teachers

In efforts to improve the performance of our nation's schools, attempts are being made to strengthen the licensing process and to create an alternative in the form of professional certification. Changes in licensing requirements have so far focused on examinations of debatable relevance and alternative licensing schemes that hold considerable promise, but that are also subject to administrative misuse. Plans for certification have the potential for creating generally accepted national standards.

• State licensing regulations should be adjusted to be consistent with reformed preservice programs.

• The various alternative ways to obtain teaching licenses (late entry, long internship, etc.) should be critically evaluated.

• Questionable routes to licensing, such as emergency certification and seniority rules that cut across disciplines, should be eliminated.

• An independent national committee composed of biologists, teachers, biology educators, and state school personnel should evaluate licensing and certification plans with emphasis on their adequacy for assessing competence to teach biology as a process of inquiry and discovery. Only a broad consensus on national standards will engage all the groups required for reform to succeed.

• The role of mentor teachers should be clarified. Mentoring is not a common working concept for today's teachers, but it is a cornerstone of any reform movement. We propose that mentor teachers be involved in developing appropriate criteria for defining mentor teachers, training new teachers, disseminating new curricular materials, and contributing to local and national curriculum development and other professional activities.

Inservice Education: How Teachers Continue to Learn

Science has a continuously changing frontier, and society today is characterized by change generated by science and technology. Moreover, science teaching is a profession, and a notable difference between professions and other occupations is that professionals are responsible for their own continuing education. As part of their professional development, teachers must engage regularly in inservice activities that update their knowledge of science and, especially in their early years, enhance their effectiveness in the classroom. Few present inservice opportunities are part of a conceptual approach that advances the dual goals of increasing understanding of a discipline and honing insight as to how it is best learned.

• There is clear need for national leadership in identifying and defining the kinds of inservice programs that will be most successful in fostering inquiry-based teaching—teaching that promotes interest, curiosity, and increasing understanding of scientific concepts. Effective inservice programs must be:

—Attractive enough to entice many teachers to participate and appropriate to teachers' needs, as identified by biology teachers, biology educators, and biologists.

—Conceptually organized, eventually operated in conjunction with preservice programs, and run on a continuing basis.

—Able to compensate teachers for their time.

—Associated with opportunities for teachers to obtain small grants to bring new approaches to the classroom.

—Constantly evaluated for their effectiveness.

—Scheduled with enough flexibility to ensure attendance.

—Designed to combine understanding of *what* to teach with knowledge and experience of *how* to teach.

—Designed and conducted with the collaboration of experienced science teachers, educators, and research scientists.

—Coupled to mechanisms for disseminating new information throughout the school district.

• Teachers need support after the inservice work, and they must be given time to assimilate the knowledge and suggestions proposed during inservice programs, to consider how the changes will affect teaching and learning in their classrooms, and to consult with colleagues. Attention must be given to the need for long-term collaborative arrangements among industry, college and university biology and biology-education research communities, and schools. Program development should involve primarily federal and private sponsorship, rather than expressly commercial inservice ventures.

• What has been said about inservice programs for high-school teachers generally holds for elementary-school and middle-school teachers. The available models are fewer and the backgrounds, motivations, and interests of the teachers are different, even though the challenge is just as important. Because the nature of the task is different from that posed for high-school teachers, we recommend the development of distinct cooperative inservice programs for these cadres of teachers as well.

• We need improved mechanisms for assessing the success of the various current experiments in inservice education; the assessments must be more sophisticated than the traditional recourse to average scores of students on regional or national examinations. We need improved models for distributing information developed in successful inservice programs and for engaging the participation of additional teachers. These are not matters that can be attended to in a single effort; continuous evaluation involving longitudinal and case studies will be required.

Recruiting Scientists, Teachers, Technicians, and Physicians

Teaching science as a process of knowing about the natural world is appropriate for all students, whether or not they will be professionally engaged in science or teaching. But an adequate supply of scientists and engineers is also necessary for the nation's survival. Schools can best meet the challenge by ensuring that all students are excited by science in their classrooms. Moreover, universities must encourage their science departments to provide better science training to both undergraduate science students and prospective precollege science teachers. For example:

• University leaders must make it clear to their science departments that the quality and quantity of the service that each department provides to precollege science teachers (both preservice and inservice training) and to the general education of nonscience majors will be an important consideration in the distribution of university resources and faculty positions.

• Universities should develop programs that integrate all interested local precollege science teachers into the various science communities of the university.

• Major universities should be expected to develop permanent summer inservice institutes for precollege science teachers either developed locally or based on successful model institutes held elsewhere. Outstanding faculty from university science departments should be recruited to teach in these institutes, side by side with outstanding precollege mentor teachers.

The reform of education for biology teachers, particularly at the preservice and certification levels, must address the growing disparity between the number of minority-group students and the number of minority-group teachers. Moreover, programs for educating teachers must compete with other attractive professions. Current efforts to restructure the teaching profession—which ought to result in greater professionalism, higher salaries, better working conditions, and more appropriate assessment techniques—will all make teaching more attractive to able minority-group students.

• One mechanism for recruiting is a national fellowship program for science teachers. Such a program could have several features that would further other, collateral goals. A competitive fellowship program could attract some of the ablest biology or elementary-education majors to science teaching. It could be of immense help in reaching minority groups, which are now underrepresented in the teaching force. With foundation underwriting, it would be possible to couple the use of fellowships to institutions that have shown interest and imagination in addressing the kinds of changes that are required in preservice education. Prestigious fellows who have studied at institutions that created the best conditions for research have made some of the most important contributions to science, and a corresponding formula needs to be tried in education. Fellowships need not be exclusively for future high-school teachers; they could be used to attract individuals with an interest in science to teach in elementary and middle schools. Similar fellowships could be offered to established teachers; in the most ambitious form, they might be used to underwrite year-long sabbaticals, during which teachers would attend universities and participate in the development of new preservice and inservice programs, as well as improve their own knowledge of science and how students learn it.

• Practices that discourage females and minority-group members from achieving their full potential in science and mathematics must be identified and eliminated.

• Colleges and universities should actively recruit women and minority-group members to careers in science and science teaching. To that end, stronger links could be forged between the historically black colleges and graduate and professional schools in research universities. Community colleges are another source of potential talent that has not been fully tapped.

• Gifted students or those who are excited by biology do not require a conceptually different curriculum from other students. The needs of all students will be best met by seeing that classrooms are staffed with able teachers who have a deep understanding of fundamental biological concepts, enjoy teaching science as a process of discovery, and are flexible and creative in addressing the needs of individual students. All the recommendations in this report are directed to that goal, and therefore a future supply of outstanding science teachers is required.

• Community-based activities that foster involvement of parents should be encouraged and extended.

• Early awareness of the wide array of vocational opportunities that can build on an interest in biology should raise the expectations of young people for success after graduation, particularly women, minority-group members, and students who lack sufficient parental guidance or other environmental stimuli. To ensure that advice does not result in frustration and disillusionment in a restricted job market, the information provided to students must be current, must be related to local conditions, and, wherever possible, must engage the cooperation of potential employers.

Integrating Biology with Other Sciences

• Models for integrated or parallel programs in biology, chemistry, physics, and mathematics should be developed and supported for both high schools and lower schools.

Special Science Schools and Centers

• The relative autonomy of both state-sponsored residential schools for science and mathematics and centers for science and technology provides a unique opportunity for these institutions to serve as "laboratories" for curricular reform. In addition to providing high-quality instruction, they should be encouraged to continue in the development of new curricula, instructional materials, and techniques for assessment. They can also serve as inservice centers for teachers from local high schools. For such experiments to have maximal impact nationally, mechanisms should be devised for comparing and assessing the programs at the several schools and centers and for disseminating the results broadly in the educational community.

• Research is required to assess the effects of magnet schools on the students they serve and on the associated neighborhood schools.

Leadership

In Chapter 8, we propose that the National Academy of Sciences through the National Research Council assume substantial responsibility for lighting the path to better science education for all by creating a standing body charged with tracking the health of science education in the nation. Creating a board that will be able to operate effectively will not be a simple task, however. Unlike the mathematics community, the scientific community is fragmented into many disciplines that rarely discuss with each other questions of either instruction or curriculum. While stressing the urgent need for action, we do not underestimate the obstacles. But we see a board within the NRC as offering an especially promising opportunity for building bridges between the science and education communities—bridges that will facilitate the kind of broad consensus that is essential if we are to achieve quality education in science.

Such a body would have no legal authority to lead; it would have to show the way by displaying its vision of the future, by emanating "the intellectual energy to persuade the community or the culture of the wisdom and validity of the vision," and by making "the vision practicable, and compelling." Moreover, it must engage the participation of outstanding teachers and science educators.

We would like to think that this report makes an important contribution to defining that vision, but our efforts will achieve little unless this foundation is built on. The issues that need to be addressed continuously—that provide the agenda for the kind of body that we propose be created within the National Research Council—embrace aspects of our earlier recommendations and are spelled out in detail in Chapter 8. The agenda is in fact open-ended, but includes the following:

- Developing recommendations for a science curriculum that starts in the early grades and builds in a coherent way through high school and into college.
- Developing standards for the quality of textbooks; providing critical and thorough examination for their accuracy, readability, coherence, and effectiveness in conveying science as a process and as a way of knowing; and guiding the selection of textbooks in states and school districts.
- Evaluating the role of national and state examinations.
- Creating criteria for evaluating the effectiveness of preservice programs for teacher education, stressing the linkage of pedagogy and content.
- Promoting interdisciplinary cooperation in the development of science curricula, the use of laboratories, and the preparation of teachers.
- Developing standards and criteria for inservice programs based on educational research, guiding the creation of new programs, evaluating the effectiveness of programs, and creating mechanisms for the wide dissemination of successful models.
- Finding new ways to promote professionalism in the community of teachers.
- Identifying research needs in science education.
- Creating and enhancing mechanisms for the collection and dissemination of information on science education, perhaps including computer-based networks or even regional institutions that teachers could visit to obtain experience with new materials and laboratory activities.
- Finding new ways to interest women and ethnically diverse students in careers in science and teaching and assessing more effective ways to teach them.
- Stimulating wider appreciation for the role of science in society.

Although the committee did not analyze in depth the costs of its recommendations, it is clear that a major commitment of funds will be needed to realize the goals set out in this report. The scientific community can provide advice about the distribution of funds to the various areas that need to be addressed to improve science education. Through an on-going collaboration among scientists, teachers, and science educators, such a board would constitute a unique forum for identifying areas needing the most immediate intervention and for recommending and evaluating activities responsive to those needs.

In summary, implementation of plans to reform science education requires national leadership from the scientific community. National leadership must develop national consensus that will press the key players into action—to produce well-educated teachers, to insist on fine instructional materials, to identify outstanding model curricula and make them available to all schools, and to encourage cooperation among students, parents, and other adults in the process of education. With this vision before us, we can approach the end of the century with new confidence in our educational system.

References

AAAS (American Association for the Advancement of Science). 1988. Science Books and Films 23(4) (March/April).

AAAS (American Association for the Advancement of Science). 1989. Project 2061. Science for All Americans. Washington, D.C.: AAAS.

Airasian, P. W. Institutional barriers to school change, pp. 252-265. In W. G. Rosen, Ed. High-School Biology Today and Tomorrow. Washington, D.C.: National Academy Press.

Aldridge, B. G. 1989. Fire up secondary school science. School Administrator 46(7):18-20.

Alston, D. A. 1988. Recruiting Minority Classroom Teachers: A National Challenge. Washington, D.C.: National Governors' Associaton.

Anderson, C. W. 1989. Assessing student understanding of biological concepts, pp. 55-70. In W. G. Rosen, Ed. High-School Biology Today and Tomorrow. Washington, D.C.: National Academy Press.

Anderson, O. R. 1989. The Teaching and Learning of Biology in the United States. Second IEA Science Study (IEA). New York, N.Y.: Teachers College, Columbia University.

Arons, A. B. 1989. What science should we teach?, pp. 13-20. In Curriculum Development for the Year 2000. A BSCS Thirtieth Anniversary Symposium. Colorado Springs, Colo.: Biological Sciences Curriculum Study.

Atkinson, R. C. 1989. Supply and demand for science and engineering PhDs: A national crisis in the making. Remarks to the Regents of the University of California, February 16, 1989.

Bethel, L. J. 1984. Science teacher preparation and professional development, pp. 143-157. In D. Holdzkom and P. B. Lutz, Eds. Research Within Reach: Science Education. Washington, D.C.: National Science Teachers Association.

Blank, R. K. 1988. The role of state policies in improving science education, pp. 61-96. In A. B. Champagne and I. M. Baden, Eds. Science Teaching: Making the System Work. Washington, D.C.: American Association for the Advancement of Science.

Blank, R. K. 1989. Educational Effects of Magnet High Schools. Paper presented at the Conference on Choice and Control in American Education, University of Wisconsin-Madison, May 17-19, 1989. Madison, Wis.: National Center on Effective Secondary Schools.

Blumberg, F., M. Epstein, W. MacDonald, and I. Mullis. 1986. A Pilot Study of Higher-Order Thinking Skills Assessment Techniques in Science and Mathematics. Final Report, Parts I and II. National Assessment of Educational Progress. Princeton, N.J.: Educational Testing Service.

Blystone, R. V. 1989. Biology learning based on illustrations, pp. 155-164. In W. G. Rosen, Ed. High-School Biology Today and Tomorrow. Washington, D.C.: National Academy Press.

Brandwein, P. F. and A. H. Passow, Eds. 1989. Gifted Young in Science: Potential Through Performance. Washington, D.C.: National Science Teachers Association.

BSCS (Biological Sciences Curriculum Study). 1989. Science for middle school students. BSCS. The Natural Selection. 1958-1989. September: 5.

Carnegie Council on Adolescent Development. 1989. Turning Points: Preparing American Youth for the 21st Century. Washington, D.C.: Carnegie Corporation of New York.

Carnegie Task Force on Teaching as a Profession. 1986. A Nation Prepared: Teachers for the 21st Century. Washington, D.C.: Carnegie Forum on Education and the Economy.

Carnevale, A. P., L. J. Gainer, and A. S. Meltzer. 1988. Workplace Basics: The Skills Employers Want. Alexandria, Va.: American Society for Training and Development.

Champagne, A. B., and I. M. Baden. 1988. Science teaching: Making the system work, pp. 1-39. In A. B. Champagne and I. M. Baden, Eds. Science Teaching: Making the System Work. Washington, D.C.: American Association for the Advancement of Science.

Cho, H.-H., J. B. Kahle, and F. H. Nordland. 1985. An investigation of high school biology textbooks as sources of misconceptions and difficulties in genetics and some suggestions for teaching genetics. Science Education 69(5):707-719.

Chubb, J. E. 1988. Why the current wave of education reform will fail. Public Interest 90(Winter): 28-49.

Clark, G. M., Ed. 1969. Life Sciences in the Middle School. Biological Sciences Curriculum Study, Special Publication No. 7. Boulder, Colo: University of Colorado.

College Board. 1987. Advanced Placement Course Description: Biology. New York: The College Entrance Examination Board.

Committee on Policy for Racial Justice. 1989. Visions of a Better Way: A Black Appraisal of Public Schooling. Washington, D.C.: Joint Center for Political Studies Press.

Conant, J. B. 1960. Recommendations for Education in the Junior High School Years. Princeton, N.J.: Educational Testing Service.

Connecticut Department of Education. 1988a. BEST: Beginning Educator Support and Training Program. Hartford, Conn.: Connecticut Department of Education.

Connecticut Department of Education. 1988b. Connecticut Continuum: Connecticut's Commitment to the Teaching Profession. Hartford, Conn.: Connecticut Department of Education.

Connecticut Department of Education. 1988c. Cooperating Teacher Program. Hartford, Conn.: Connecticut Department of Education.

Cooperman, S., and L. Klagholz. 1985. New Jersey's alternative route to certification. Phi Delta Kappan 66(10):691-695.

Cronin, J. E., and A. J. Almquist. 1988. Fact, fancy, and myth on human evolution. Current Anthropology 29(3):520-522.

Crooks, T. J. 1988. The impact of classroom evaluation practices of students. Review of Educational Research 58(4):438-481.

Driver, R., D. Child, R. Gott, J. Head, S. Johnson, C. Worsley, and F. Wylie. 1964. Science in schools: Age 15. Report No. 2. Assessment of Performance Unit, University of Leeds. Publication available from: Association for Science Education, College Lane, Hatfield, Hertfordshire, England, AL 10 9AA.

Education Commission of the States. 1985. Report of the Business Advisory Committee. Denver, Colo.: Education Commission of the States.

Eichhorn, D. H. 1966. The Middle School. New York, N.Y.: Center for Applied Research in Education, Inc.

Eisenhardt, W. B. 1976. A search for the predominant causal sequence in the interrelationship of interest in academic subjects and academic achievement. A cross-lagged panel correlation study. Unpublished doctoral dissertation, Duke University. Xerox University Microfilms, Ann Arbor, Michigan.

Epstein, K. C. 1987. Many Ohio science teachers favor study of creationism. Plain Dealer (Cleveland, Ohio), September 3.

Eylon, B.-S., and M. C. Linn. 1988. Learning and instruction: An examination of four research perspectives in science education. Review of Educational Research 58(3):251-301.

Frank, P. 1957. Philosophy of Science: The Link Between Science and Philosophy. Englewood Cliffs, N.J.: Prentice Hall.

Gallagher, J. 1986. A summary of research in science education—1985. Science Education 71:271-455.

GAO (U.S. General Accounting Office). 1984. New Directions for Federal Programs to Aid Mathematics and Science Teaching. (GAO/PEMD-84-5) Washington, D.C.: U.S. General Accounting Office.

GAO (U.S. General Accounting Office). 1988. R&D Funding: The Department of Education in Perspective. (GAO/PEMD-88-18FS). Washington, D.C.: U.S. General Accounting Office.

Giamatti, A. B. 1988. A Free and Ordered Space: The Real World of the University. New York, N.Y.: W. W. Norton and Co.

Gould, S. J. 1988. The case of the creeping fox terrier clone. Natural History January:16-24.

Green, J. 1987. The Next Wave: A Synopsis of Recent Education Reform Reports. Denver, Colo.: Education Commission of the States.

Green, K. C. 1989. A Profile of Undergraduates in the Sciences. Keynote presentation, meeting of the Sigma Xi National Advisory Group on Undergraduate Education in the Sciences, Mathematics and Engineering at the Wingspread Conference Center, Racine, Wis., on January 23, 1989.

Guskey, T. R. 1986. Staff development and the process of teachers change. Education Research May:5-12

Guyton, E., and E. Farokhi. 1987. Relationships among academic performance, basic skills, subject matter knowledge, and teaching skills of teacher education graduates. Journal of Teacher Education 38(5):37-52.

Haney, W. 1984. Testing reasoning and reasoning about testing. Review of Educational Research 54(4):597-654.

Hofstein, A., and V. N. Lunetta. 1982. The role of laboratory in science teaching: Neglected aspects of research. Review of Educational Research 52(2): 201-217.

Holmes Group. 1986. Tomorrow's Teachers: A Report of the Holmes Group. East Lansing, Mich.: College of Education, Michigan State University.

Holt, Rinehart and Winston. 1989. Modern Biology. New York, N.Y.: Holt, Rinehart and Winston.

Hooper, C. 1990. NSF's $251 million catalyst for change. Journal of NIH Research 2(April):28-29.

Hummel, T. J., and S. M. Strom. 1987. The relationship between teaching experience and satisfaction with teacher preparation: A summary of three surveys. Journal of Teacher Education 38(5):28-36.

Hurd, P. D. 1989a. A life science core for early adolescents. Middle Schools Journal 20(5):20-24.

Hurd, P. D. 1989b. Problems and issues in science-curriculum reform and implementation, pp. 291-297. In W. G. Rosen, Ed. High-School Biology Today and Tomorrow. Washington, D.C.: National Academy Press.

IEA (International Association for the Evaluation of Educational Achievement). 1988. Science Achievement in Seventeen Countries. A Preliminary Report. New York, N.Y.: Pergamon Press.

Institute of Medicine. 1989. Research on Children and Adolescents with Mental, Behavioral and Developmental Disorders. Washington, D.C.: National Academy Press.

Jackson, P. W. 1983. The Reform of Science Education: A Cautionary Tale. Daedalus Spring:143-166.

Kahle, J. B. 1982. Can positive minority attitudes lead to achievement gains in science? Analysis of the 1977 National Assessment of Educational Progress, Attitudes Toward Science. Science Education 66(4):539-546.

Kahle, J. B. 1985. Retention of Girls in Science: Case Studies of Secondary Teachers. In J. B. Kahle, Ed., Women in Science: A Report From the Field. Philadelphia, Pa.: The Falmer Press.

Kahle, J. B. 1987. Scores: A project for change? International Journal of Science Education 9(3):325-333.

Kahle, J. B., and M. K. Lakes. 1983. The myth of equality in science classrooms. Journal of Research in Science Teaching 20:131-140.

Kennedy, D. 1990. Stanford in Its Second Century. Stanford University Campus Report, April 11, pp. 17-18.

Klein, S. P., and J. Kosecoff. 1973. Issues and Procedures in the Development of Criterion Referenced Tests. TM Report 26. ERIC Clearinghouse on Tests, Measurement, and Evaluation, Educational Testing Service, Princeton, New Jersey.

Lapointe, A. E., N. A. Mead, and G. W. Phillips. 1989. A World of Differences. An International Assessment of Mathematics and Science. Princeton, N.J.: Educational Testing Service.

Leonard, W. H., G. R. Cavana, and L. F. Lowery. 1981. An experimental test of an extended discretion approach for high school laboratory investigations. Journal of Research in Science Teaching 18:497-507.

Lerner, L. S., and W. J. Bennetta. 1988. The treatment of theory in textbooks. Science Teacher. April:37-41.

Linn, M. C. 1986. Science. In R.F. Dillon, and R. J. Sternberg. Cognition and Instruction. New York, N.Y.: Academic Press.

Linn, M. C., and J. S. Hyde. 1989. Gender, mathematics, and science. Educational Researcher 18(8):17-27.

MacDonald, W. B. 1989. The advanced-placement biology examination: Its rationale, development, structure, and results, pp. 71-78. In W. G. Rosen, Ed. High-School Biology Today and Tomorrow. Washington, D.C.: National Academy Press.

Mayer, W. V. 1978. Biology Teachers' Handbook. 3rd ed. New York, N.Y.: John Wiley and Sons.

Mayer, W.V. 1986. Biology Education in the United States During the Twentieth Century. The Quarterly Review of Biology 61(4):481-507.

Merriam, R. W. 1988. A function in trouble: Undergraduate science teaching in research universitites. Journal of College Science Teaching November:102, 106.

Miller, J. D. 1989. The development of interest in science, pp. 79-90. In W. G. Rosen, Ed. High-School Biology Today and Tomorrow. Washington, D.C.: National Academy Press.

Moyer, W. A. 1989. Developing a synthesis between seventh-grade life-science and tenth-grade biology, pp. 131-138. In W. G. Rosen, Ed. High-School Biology Today and Tomorrow. Washington, D.C.: National Academy Press.

Mullis, I. V. S. 1989. What high-school juniors know about biology: Perspectives from NAEP, the Nation's Report Card, pp. 91-98. In W. R. Rosen, Ed. High-School Biology Today and Tomorrow. Washington, D.C.: National Academy Press.

Mullis, I. V. S., and L. B. Jenkins. 1988. The 1986 Science Report Card: Elements of Risk and Recovery. National Assessment of Educational Progress, Educational Testing Service. Princeton, N.J.: Educational Testing Service.

Murname, R. J., and S. A. Raizen. 1988. Improving Indicators of the Quality of Science and Mathematics Education in Grades K-12. Committee on Indicators of Precollege Science and Mathematics Education, National Research Council. Washington, D.C.: National Academy Press.

NABT (National Association of Biology Teachers). 1985. NABT Teaching Standards. News & Views. Reston, Va.: National Association of Biology Teachers.

Nachmias, R., and M. C. Linn. 1987. Evaluations of science laboratory data: The role of computer-presented information. Journal of Research in Science Teaching 24(5):491-506.

National Academy of Sciences. 1984. High schools and the Changing Workplace: The Employer's View. Report of the Panel on Secondary School Education for the Changing Workplace, Committee on Science, Engineering and Public Policy. Washington, D.C.: National Academy Press.

National Commission on Excellence in Education. 1983. A Nation at Risk: The Imperative for Educational Reform. Washington, D.C.: U.S. Department of Education.

National Governors' Association Center for Policy Research. 1988. Getting Ready for the National Board for Professional Teaching Standards. Washington, D.C.: National Governors' Association.

National Governors' Association. 1990. National Education Goals. Adopted by the members of the National Governors' Association on February 25, 1990. (Unpublished manuscript)

NCRTE (National Center for Research on Teacher Education). 1988. Teacher education and learning to teach: A research agenda. Journal of Teacher Education 39(6):27-32.

New Jersey State Department of Education. 1989. The Provisional Teacher Program. Fifth Year Report. Trenton, N.J.: New Jersey State Department of Education.

New York Times. New Jersey, in 5 years, solves teacher shortage. September 13, 1989. B10.

New York Times. Japan keeps up the big spending to maintain its industrial might. April 11, 1990. A1, D7.

Novak, J. D. 1988. Learning science and the science of learning. Studies in Science Education 15:77-101.

NRC (National Research Council), Mathematical Sciences Education Board. 1989a.
    Everybody Counts: A Report to the Nation on the Future of Mathematics
    Education. Washington, D.C.: National Academy Press.
NRC (National Research Council). 1989b. A Common Destiny: Blacks and American
    Society. Washington, D.C.: National Academy Press.
NSF (National Science Foundation). 1988. Program Solicitation. Programs for Middle
    School Science Instruction. Washington, D.C. : National Science Foundation.
NSRC (National Science Resources Center). 1988. Science for Children: Resources
    for Teachers. Washington, D.C.: National Academy Press.
NSTA (National Science Teachers Association). 1984. Standards for the Preparation and
    Certification of Teachers of Science, K-12. Washington, D.C.: National Science
    Teachers Association.
OTA (U.S. Congress, Office of Technology Assessment). 1988. Elementary & Sec-
    ondary Education for Science and Engineering (OTA-SET-41). Washington, D.C.:
    U.S. Government Printing Office.
Paul, D. B. 1987. The nine lives of discredited data: Old textbooks never die—they
    just get paraphrased. Sciences May/June:26-30.
Pauling, L. 1939. The Nature of the Chemical Bond and the Structure of Molecules
    and Crystals. Ithaca, N. Y.: Cornell University Press.
Resnick, L. B. 1987. Education and Learning to Think. Report of the Committee
    on Mathematics, Science, and Technology Education, National Research Council.
    Washington, D.C.: National Academy Press.
Resnick, D. P., and L. B. Resnick. 1985. Standards, curriculum, and performance: A
    historical and comparative perspective. Educational Researcher 14:5-21.
Richter, B., and D. Wenzel. 1986, 1987. The Museum of Science and Industry Basic
    List of Children's Science Books 1986 and 1987. Chicago, Ill.: American Library
    Association.
Robinson, J. T. 1979. A critical look at grading and evaluating practices, pp. 1- 31. In
    M. B. Rowe, Ed. What Research Says to the Science Teacher. Vol 2. Washington,
    D.C.: National Science Teachers Association.
Robinson, J. T. 1989. Issues in objectives and evaluation, pp. 45-54. In W. G. Rosen,
    Ed. High-School Biology Today and Tomorrow. Washington, D.C.: National
    Academy Press.
Science Curriculum Framework and Criteria Committee. 1984. Science Framework
    Addendum for California Public Schools: Kindergarten and Grades One Through
    Twelve. Sacramento, Calif.: California State Department of Education.
Smith, E. L., and C. W. Anderson. 1984. Plants as producers: A case study of
    elementary science teaching. Journal of Research in Science Teaching 21(7):685-
    698.
Stake, R. E., and J. A. Easley. 1978. Case Studies in Science Education Vol. II.
    Design, Overview and General Findings. Center for Instructional Research and
    Curriculum Evaluation and Committee on Culture and Cognition, University of
    Illinois at Urbana-Champaign.
Stiggins, R. J. 1987. Design and Development of Performance Assessments. NCME
    Instructional Module. Educational Measurement: Issues and Practice. Washington,
    D.C.: National Council for Measurement in Education.
Stiggins, R. J., D. A. Frisbie, and P. A. Griswold. Undated. Inside High School
    Grading Practices: Building a Research Agenda. Portland, Ore: Northwest
    Regional Educational Laboratory.

Subcommittee on Instructional Materials and Publications, Committee on Educational Policies, Division of Biology and Agriculture, National Academy of Sciences-National Research Council. 1957. Criteria for preparation and selection of science textbooks. A.I.B.S. Bulletin November:26-28.

Tamir, P., and F. Glassman. 1970. A practical examination for BSCS students. Journal of Research in Science Teaching 7(2):197-212.

Task Force on Women, Minorities and the Handicapped in Science and Technology. 1989. Changing America: The New Face of Science and Engineering. Washington, D.C.: Task Force on Women, Minorities and the Handicapped in Science and Technology.

Taylor, G. S. 1989. Different schools: Same barriers, pp. 278-288. In W. G. Rosen, Ed. High-School Biology Today and Tomorrow. Washington, D.C.: National Academy Press.

Tobin, K. G. 1989. Learning in science classrooms, pp. 25-38. In Curriculum Development for the Year 2000. A BSCS Thirtieth Anniversary Symposium. Colorado Springs, Colo.: Biological Sciences Curriculum Study.

Tobin, K. G., and J. J. Gallagher. 1987. What happens in high school science classrooms? Journal of Curriculum Studies 19:549-560.

Tyson-Bernstein, H. 1988. A Conspiracy of Good Intentions. Washington, D.C.: Council for Basic Education.

U.S. Department of Education. 1989. Back-to-School Forecast. Washington, D.C.: U.S. Department of Education.

U.S. Department of Labor, U.S. Department of Education, and U.S. Department of Commerce. 1988. Building a Quality Workforce. Washington, D.C.: U.S. Department of Labor.

Van Til, W., G. F. Vars, and J. H. Lounsbury. 1961. Modern Education for the Junior High School Years. Indianapolis, Ind.: Bobbs Merrill Co.

Vetter, B. M. 1987. Women's Progress. Mosaic 18(1):2-9.

Vetter, B. M. 1989. Look who's coming to school. Changing demographics: Implications for science education, pp. 1-9. In Curriculum Development for the Year 2000. A BSCS Thirtieth Anniversary Symposium. Colorado Springs, Colo.: Biological Sciences Curriculum Study.

Watson, J. D. 1965. Molecular Biology of the Gene. New York, N.Y.: W. A. Benjamin, Inc.

Weiss, I. R. 1987. Report of the 1985-86 National Survey of Science and Mathematics Education. Research Triangle Park, N.C.: Research Triangle Institute.

Weiss, I. R. 1989. Science and Mathematics Education Briefing Book. Chapel Hill, N.C.: Horizon Research, Inc.

Welch, W. W., L. J. Harris, and R. E. Anderson. 1984. How many are enrolled in science? Science Teacher December:14-19.

West, L. H. T., and A. L. Pines, Eds. 1985. Cognitive Structure and Conceptual Change. New York, N.Y.: Academic Press.

Westheimer, F. H. 1987. Are our universities rotten at the "core"? Science 236:1165-1166.

Whyte, J. S. 1965. Girls into Science and Technology. London, England: Routledge and Kegan Paul.

Wilson, E. B. 1896. Cell in Development and Inheritance. New York, N.Y.: MacMillan Company.

APPENDIXES

APPENDIX A
PRINCIPLES AND GUIDELINES FOR THE USE OF ANIMALS IN PRECOLLEGE EDUCATION

Institute of Laboratory Animal Resources
Commission on Life Sciences
National Research Council
National Academy of Sciences
National Academy of Engineering

The humane study of animals in precollege education can provide important learning experiences in science and ethics and should be encouraged. Maintaining classroom pets in preschool and grade school can teach respect for other species, as well as proper animal husbandry practices. Introduction of secondary school students to animal studies in closely supervised settings can reinforce those early lessons and teach the principles of humane care and use of animals in scientific inquiry. The National Research Council recommends compliance with the following principles whenever animals are used in precollege education or in science fair projects.

PRINCIPLE 1

Observational and natural history studies that are not intrusive (that is, do not interfere with an animal's health or well-being or cause it discomfort) are encouraged for all classes of organisms. When an intrusive study of a living organism is deemed appropriate, consideration should be given first to using plants (including lower plants such as yeast and fungi) and invertebrates with no nervous systems or with primitive ones (including protozoa, planaria, and insects). Intrusive studies of invertebrates with advanced nervous systems (such as octopi) and vertebrates should be used only when lower invertebrates are not suitable and only under the conditions stated below in Principle 10.

PRINCIPLE 2

Supervision shall be provided by individuals who are knowledgeable about and experienced with the health, husbandry, care, and handling of the animal species used and who understand applicable laws, regulations, and policies.

PRINCIPLE 3

Appropriate care for animals must be provided daily, including weekends, holidays, and other times when school is not in session. This care must include

   a. nutritious food and clean, fresh water;
   b. clean housing with space and enrichment suitable for normal species behaviors; and
   c. temperature and lighting appropriate for the species.

PRINCIPLE 4

Animals should be healthy and free of diseases that can be transmitted to humans or to other animals. Veterinary care must be provided as needed.

PRINCIPLE 5

Students and teachers should report immediately to the school health authority all scratches, bites, and other injuries; allergies; or illnesses.

PRINCIPLE 6

Prior to obtaining animals for educational purposes, it is imperative that the school develop a plan for their procurement and ultimate disposition. Animals must not be captured from or released into the wild without the approval of the responsible wildlife and public health officials. When euthanasia is necessary, it should be performed in accordance with the most recent recommendations of the American Veterinary Medical Association's Panel Report on Euthanasia (*Journal of the American Veterinary Medical Association,*

188[3]:252–268, 1986, et seq.). It should be performed only by someone trained in the appropriate technique.

PRINCIPLE 7

Students shall not conduct experimental procedures on animals that

a. are likely to cause pain or discomfort or interfere with an animal's health or well-being;
b. induce nutritional deficiencies or toxicities; or
c. expose animals to microorganisms, ionizing radiation, cancer-producing agents, or any other harmful drugs or chemicals capable of causing disease, injury, or birth defects in humans or animals.

In general, procedures that cause pain in humans are considered to cause pain in other vertebrates.

PRINCIPLE 8

Experiments on avian embryos that might result in abnormal chicks or in chicks that might experience pain or discomfort shall be terminated 72 hours prior to the expected date of hatching. The eggs shall be destroyed to prevent inadvertent hatching.

PRINCIPLE 9

Behavioral conditioning studies shall not involve aversive stimuli. In studies using positive reinforcement, animals should not be deprived of water; food deprivation intervals should be appropriate for the species but should not continue longer than 24 hours.

PRINCIPLE 10

A plan for conducting an experiment with living animals must be prepared in writing and approved prior to initiating the experiment or to obtaining the animals. Proper experimental design of projects and concern for animal welfare are important learning experiences and contribute to respect for and appropriate care of animals.

The plan shall be reviewed by a committee composed of individuals who have the knowledge to understand and evaluate it and who have the authority to approve or disapprove it. The written plan should include the following:

a. a statement of the specific hypotheses or principles to be tested, illustrated, or taught;
b. a summary of what is known about the subject under study, including references;
c. a justification for the use of the species selected and consideration of why a lower vertebrate or invertebrate cannot be used; and
d. a detailed description of the methods and procedures to be used, including experimental design; data analysis; and all aspects of animal procurement, care, housing, use, and disposal.

EXCEPTIONS

Exceptions to Principles 7–10 may be granted under special circumstances by a panel appointed by the school principal or his or her designee. This panel should consist of at least three individuals, including a science teacher, a teacher of a nonscience subject, and a scientist or veterinarian who has expertise in the subject matter involved.[1] At least one panel member should not be affiliated with the school or science fair, and none should be a member of the student's family.

April 1989

[1] In situations where an appropriate scientist is not available to assist the student, the Institute of Laboratory Animal Resources (ILAR) might be able to provide referrals. Write to ILAR, National Research Council, 2101 Constitution Avenue, NW, Washington, DC 20418, or call (202)334-2590.

NABT Biology Teaching Standards

PROGRAM SCOPE: A biology education program should prepare teachers for both the junior high/middle school and senior high school levels of instruction and should be designed to educate college and university students to teach any secondary biology or other life science courses. The suggested program should include a minimum of 24 semester hours in the biological sciences, including course work to ensure the proficiencies stated below; plus a minimum of 24 semester hours in chemistry and introductory physics; and proficiency in mathematics through college algebra. A minimum of 12 semester hours in science should be upper division hours. All secondary biology teachers should be prepared to teach in at least one other science area. In addition, biology teachers must continue to improve their skills and knowledge in the ever changing world of life sciences.

Standards For Preparation of the Biology Teacher

Candidates completing a biology teacher certificate program must include the curricular goals listed below and be able to demonstrate specific skills and knowledge.

1. Knowledge of the fundamentals of Biology.

   - Demonstrate a knowledge of basic concepts and of laboratory techniques concerned with the study of: systematics; development; evolution; genetics; ethical implications of technology (recombinant DNA, organ transplant, in vitro fertilization); ecology; behavior; cell biology; bio-energetics; homeostatic mechanisms; and all the life processes in animals, plants and microbes.

2. Knowledge of the interrelationships of living organisms with their biotic and physical environments, including field experiences and the study of ecology or environmental biology.

   - Demonstrate in writing a knowledge of the basic concepts of ecological population factors; ecosystems; energy flow; nutrient cycles and the sociobiological aspects of ecology.

   - Demonstrate an ability to conduct and direct meaningful field trips and investigations concerned with obtaining information on concepts of ecological populations; ecosystems; energy flow; nutrient cycles and the sociobiological aspects of ecology.

3. Knowledge of chemistry, mathematics, and physical science or physics, and computer science —

   - Demonstrate: a basic knowledge of the concepts and a command of the laboratory techniques equivalent to those included in general college chemistry; the concepts equivalent to those included in lower division, undergraduate physical science course, or a college physics I course; a command, or working ability of mathematics equivalent to that in college algebra; and an ability to utilize computers in teaching and in record storage.

4. A methods course for biology teaching designed to organize, plan, present, and evaluate the learning of biology subject matter content.

   - Demonstrate a functional knowledge of the science inquiry processes and be able to distinguish between assumptions, hypotheses, theories, data, controls, independent and dependent variables, and generalizations.

   - Define and describe a philosophy of present-day science teaching.

   - Demonstrate a command of the mechanics of everyday teaching, including laboratory and field experiences.

- Select, purchase, operate, and maintain equipment and supplies used in teaching biology.

- Use current biology curricular materials in the classroom.

- Demonstrate an ability to develop curricula that motivate students as well as consider individual differences.

- Demonstrate an ability to construct and administer student evaluation instruments for subject matter concepts, principles, and techniques.

- Demonstrate a commitment and dedication to education of early adolescents and continual self-improvement.

- Foster enthusiasm about biology in students of diverse backgrounds.

- Demonstrate interest in professional growth by actively participating in local, regional, or national biology association programs.

Standards for Professional Growth of the Biology Teacher

Teachers who wish to maintain their skills and knowledge gained in undergraduate work must include the following goals in their profession.

1. Maintain standard of excellence and broaden knowledge of life sciences.

   - Demonstrate professionalism by participating in a biological science teacher education program which will lead to a higher degree.

   - Participate in biology inservice programs and/or summer institutes to learn new teaching methods and laboratory techniques.

   - Participate in local, regional, and national biology conferences to keep abreast of new trends and discoveries.

   - Demonstrate commitment to learning by reading professional journals.

2. Establish close relationship with scientific community, businesses, and industries.

   - Demonstrate interest in scientific community by participating in local and national biology organizations.

   - Develop communication with local businesses, nonprofit organizations and private institutions.

   - Demonstrate leadership by taking active role in maintaining scientific integrity in the community and by sharing biology teaching ideas with colleagues.

Standards for the Preparation And Certification of Secondary School Teachers of Science

I. Science Content Preparation

The program for preparing secondary school teachers of science should require specialization in one of the sciences (i.e. preparation equivalent to the bachelor's level) as well as supporting course work in other areas of science. The programs should require a minimum of 50 semester hours of course work in one or more of the sciences and additional course work in related content areas such as mathematics, statistics, and computer applications to science teaching. The programs and courses should be designed to develop a breadth of scientific literacy that will provide the preservice teacher with

• positive attitudes toward science and an accompanying motivation to be a lifelong learner in science;

• competency in using the processes of science common to all scientific disciplines, including the skills of investigating scientific phenomena, interpreting the findings, and communicating results;

• competency in a broad range of research, laboratory and field skills;

• knowledge of scientific concepts and principles and their applications in technology and society;

• an understanding of the relationship between science, technology, society and human values; and

• decision-making and value-analysis skills for use in solving science-related problems in society.

Overall, the programs should be designed for the unique needs of secondary school science teachers.

II. Science Teaching Preparation

Science Teaching Methods and Curricula

The program should prepare preservice teachers in the methods and curricula of science. Method courses should model desired teaching behavior in the secondary classroom. These experiences should develop a wide variety of skills, including those which help preservice science teachers to

• teach science processes, attitudes, and content to learners with a wide range of abilities and socio-economic and ethnic backgrounds;

• become knowledgeable of a broad range of secondary school science curricula, instructional strategies and materials, as well as how to select those best suited for a given teaching and learning situation;

• become proficient in constructing and using a broad variety of science evaluation tools and strategies; and

• become knowledgeable about the learning process, how people learn science, and how related research findings can be applied for more effective science teaching.

The program should include *at least* one separate course (3-5 semester hours), and preferably more, in science teaching methods and curricula.

Communication Skills and Classroom Management Techniques

The program should prepare preservice teachers to speak and write effectively and demonstrate effective

NATIONAL SCIENCE TEACHERS ASSOCIATION, 1742 CONNECTICUT AVENUE, N.W., WASHINGTON, D.C. 20009 use of classroom management techniques when teaching laboratory activities, leading class discussions, conducting field trips, and carrying out daily classroom instruction in science.

Preparation in Research Skills

The program should prepare preservice teachers to conduct or apply, understand, and interpret science education research and to communicate information about such research to others (e.g., students, teachers and parents).

Safety in Science Teaching

The program should require experiences that develop the ability to identify, establish, and maintain the highest level of safety in classrooms, stockrooms, laboratories, and other areas used for science instruction.

Other Educational Experiences

Courses in other educational areas, including general curricula and methods, educational psychology, foundations and the special needs of exceptional students, should be a part of the program in order to complement the science education components described above.

III. Classroom Experience

Field Experience

Field experiences in secondary school science classrooms are essential for the thorough preparation of preservice teachers of science. The field experience of preservice teachers should begin early with an emphasis on observation, participation, and tutoring, and should progress from small to large group instruction.

The Student Teaching Experience

The student teaching experience should be full-time for a minimum of 10 weeks. The program should require student teaching at more than one educational level (such as junior high school experience combined with that of working in the high school) or in more than one area of science (i.e., biology and chemistry) if certification is sought in more than one area. The program should give prospective teachers experience with a full range of in-school activities and responsibilities.

Day-to-day supervision of the student teacher should be done by an experienced, master science teacher(s). University supervision should be provided by a person having significant secondary school science teaching experience. Responsibility for working with student teachers should be given only to highly qualified, committed individuals, and close and continuing cooperation between school and university is imperative.

IV. Supportive Preparation in Mathematics, Statistics, and Computer Use

The program should require competencies in

• mathematics as specified for each discipline;

• scientific and educational use and interpretation of statistics; and

• computer applications to science teaching, emphasizing computer tools such as: (a) computation, (b) interfacing with lab experiences and equipment, (c) processing information, (d) testing and creating models, and (e) describing processes, procedures, and algorithms.

NATIONAL SCIENCE TEACHERS ASSOCIATION,
1742 CONNECTICUT AVENUE, N.W., WASHINGTON, D.C. 20009

Standards for Each Secondary Discipline

Biology

I. The program in biology should require broad study and experiences with living organisms. These studies should include use of experimental methods of inquiry in the laboratory and field and applications of biology to technology and society.

II. The program would require a minimum of 32 semester hours of study in biology to include at least the equivalent of three semester hours in each of the following: zoology, botany, physiology, genetics, ecology, microbiology, cell biology/biochemistry, and evolution; interrelationships among these areas should be emphasized throughout.

III. The program should require a minimum of 16 semester hours of study in chemistry, physics, and earth science emphasizing their relationships to biology.

IV. The program should require the study of mathematics, at least to the pre-calculus level.

V. The program of study for preservice biology teachers should provide opportunities for studying the interaction of biology and technology and the ethical and human implications of such developments as genetic screening and engineering, cloning, and human organ transplantation.

VI. The program should require experiences in designing, developing, and evaluating laboratory and field instructional activities, and in using special skills and techniques with equipment, facilities, and specimens that support and enhance curricula and instruction in biology.

NATIONAL SCIENCE TEACHERS ASSOCIATION,
1742 CONNECTICUT AVENUE, N.W., WASHINGTON, D.C. 20009

APPENDIX D
STATE ALTERNATIVE CERTIFICATION PROGRAMS*
(AS OF JUNE 1987)

State	Program characteristics
Alabama	Certified teacher may apply for temporary math permit, 6 semesters of math yearly for 4 years to complete certification.
Arizona	Associate Teacher Authorization, renewable twice, allows holder to teach part-time or one-half year full-time under supervision of certified teacher.
California	Pass CBEST and NTE, mentor teacher and program to work for certification after 2 years.
Connecticut	Six weeks' education training and classroom experience under 90-day provisional certificate, then enter 1 year new teacher support and assessment program.
Florida	On secondary level, 3 week intensive training program and 1 year internship.
Georgia	In areas of critical need, a science/mathematics major, teacher certification test, performance based assessment, course work and 1 year internship.
Indiana	Limited license with 15 semester hours science/mathematics, renewable with 6 hours credit each year toward licensure requirements.
Louisiana	Louisiana State University program; pass NTE, course work, internship.
Maryland	Proposed "alternative programs" approved under Creative Initiatives in Teacher Education.
Mississippi	On secondary level, provisional certificate until pass NTE, complete 12 hours credit and on-the-job competency.
New Hampshire	Pass teacher certification exam, 3 year internship with master teacher.
New Jersey	Pass NTE and subject area test, employment with a training program.
New Mexico	Pass NTE, employment and individualized professional plan.
North Carolina	Lateral entry in area of critical need, employment and examination of qualifications.

Ohio	Proposed "alternative programs" approved under Flexible and Innovative Individualized Program Standard.
Oklahoma	Pass teacher certification test, employment and teacher education program.
Oregon	Certified teachers may take NTE subject exam for science/mathematics endorsement.
Pennsylvania	May teach if participating in intern program at institution of higher education.
South Carolina	In area of critical need, summer courses, training workshops in school year and 3 graduate courses within 3 years.
Tennessee	Fifth year program for second career persons with part-time classroom teaching.
Texas	In area of critical need, pass teacher certification test, course work and 1 year internship with appraisal.
Vermont	Peer Review.
Virginia	On secondary level, pass NTEV, employment and 2 years to complete 9 credit hours.
West Virginia	In area of critical need, summer session and 1 year internship.

Notes: *Programs designed, at least partially, to increase the number of science or mathematics teachers in the state.

Alternative certificate program: Teacher preparation program that enrolls uncertified individuals who hold at least a bachelor's degree.

Offers shortcuts, special assistance, or unique curricula leading to eligibility for a standard teaching credential (Adelman, 1986).

(From Blank, R. K. 1988. The role of state policies in improving science education, pp. 61-96. In A. B. Champagne and I. M. Baden, Eds. Science Teaching: Making the System Work. Washington, D.C.: American Association for the Advancement of Science. Pp. 73 and 74.)

APPENDIX E
SPECIAL SCIENCE PROGRAMS AND SCHOOLS

Some Specialized Public High Schools and Magnet High Schools of Science and Mathematics

Illinois

Whitney M. Young Magnet High School, Chicago

Maryland

Baltimore Polytechnic Institute, Baltimore
Eleanor Roosevelt High School, Greenbelt
Montgomery Blair High School, Silver Spring
Oxon Hill High School, Oxon Hill

New York

Bronx High School of Science, New York
Brooklyn Technical High School, New York
Stuyvesant High School, New York

Pennsylvania

Central High School of Philadelphia, Philadelphia
George Washington High School of Engineering and Science, Philadelphia

Texas

Science Academy of Austin, Austin

Virginia

Central Virginia Governor's School for Science and Technology, Lynchburg
Thomas Jefferson High School for Science and Technology, Alexandria

State-Sponsored Residential Schools of Science and Mathematics

The Illinois Science and Mathematics Academy (IMSA) was founded in 1986. It accepts students after their freshman year. One of its goals is "to serve as a laboratory for the development of testing, and dissemination of innovative techniques in mathematics, science and the humanities which can become a resource for secondary school teachers in Illinois and the nation." IMSA is in residential Aurora, Ill., along the high-technology corridor that includes Fermi National Accelerator Laboratory, Amoco, NALCO, AT&T Bell Laboratories, Argonne National Laboratory, and many other research laboratories.

The Louisiana School for Science, Mathematics and the Arts admitted its first class in 1983. As an extension of the public-school system, it offers a specialized program for juniors and seniors on the campus of Northwestern

University in Natchitoches, La. The school also serves as a resource center for inservice training and research or education of gifted and talented students.

The Mississippi School for Math and Science, in its second year, is in Columbus, Miss., on the campus of Mississippi University for Women. The purpose of the school "shall be to educate the gifted and talented students of the state." Students are admitted after their sophomore year; the application process is competitive.

The North Carolina School for Science and Mathematics was opened in 1980 as the nation's first public, residential high school for students with special aptitude and interest in science and mathematics. Students are accepted after their sophomore year. A recent survey found that 80% of its graduates went on to science and engineering majors in college, and two-thirds of its graduates have elected to go to college in North Carolina. It is in Durham, in the Research Triangle area.

The South Carolina Governor's School for Science and Mathematics is in its third year. It is housed in a rural setting on the campus of Coker College, in Hartsville, S.C. It is a 2-year institution that draws students from large and small high schools throughout the state. The school's goal is "to teach the students how to think, analyze and synthesize information and understand the complexities of problem solving in any discipline." The school plans to become involved in inservice teacher training by sponsoring summer institutes.

The Texas Academy of Science and Mathematics, housed at the University of North Texas, is in its second year. It offers an early-admission program that allows students who are particularly talented in science and mathematics to take their last 2 years of high school and first 2 years of college concurrently in residence on a college campus.

Local and Regional Science and Technology Centers

Roanoke Valley Governor's School, Roanoke, Va., opened in August 1985. The Governor's School serves eight school districts and 15 high schools. It offers a 3-year science and mathematics curriculum that provides accelerated opportunities for highly motivated secondary-school students.

The New Horizons Governor's School for Science and Technology, Hampton, Va., students attend classes at both New Horizons and their home school. Students also serve a mentorship in a professional or research setting as an after-school or weekend activity.

The Kalamazoo Area Mathematics and Science Center, Kalamazoo, Mich., was conceived in 1981 by the Upjohn Company and developed with cooperation of the schools of the greater Kalamazoo area. The center opened its doors in 1986. In 1989, it had a 4-year program with 400 students. The center is under control of the schools, but draws on the resources and counsel of private industry. Students spend half-days at the center, and return to home schools for the remainder of the day. The goal is to plan and deliver professional development programs for mathematics and science educators, in concert with area scientists.

Additional Science and Technology Enhancement Programs

Teacher Enhancement:

Ball State University, Indiana. A 4-week workshop in human genetics, sponsored by the National Science Foundation (NSF), couples information about content with strategies for teaching new materials. It has been oversubscribed for years, and it has fulfilled one of the prime objectives of earlier NSF-sponsored summer institutes: promoting a feeling of community among biology teachers.

DNA Literacy Program, Cold Spring Harbor Laboratory. The DNA Learning Center has developed a curriculum that centers around nine experiments culminating in the analysis of recombinant DNA. Teachers participate in a 5-day summer workshop, with a weekend followup during the winter, designed to help instructors to set up laboratory programs in their own schools. The center, in Cold Spring Harbor, N.Y., provides an interactive environment for students, teachers, and the public. Middle-school and high-school students also participate in laboratory activities at the center during the school year.

Howard Hughes Medical Institute (HHMI). As part of the outreach program of its undergraduate initiative, HHMI has given grants to 44 undergraduate institutions to improve the quality of curricula and teaching of high-school biology and related sciences. The 44 institutions are to "offer an array of academic training programs in science and mathematics for teachers and students at the junior, high school, and junior college levels."

Lawrence Livermore Laboratory. This federally funded facility in Oakland, Calif., offers a program to address shortcomings of science instruction at the elementary level and the underrepresentation of minority groups in technical fields. Each summer, scientists offer lesson workshops for teachers. Key elements include instruction by laboratory scientists, engineers, and technologists; instruction in lesson-plan preparation; and experiments with inexpensive materials that demonstrate basic scientific concepts. For many elementary-school teachers, this is the first exposure to physics or chemistry. The program strives to reduce teacher anxiety about discussing physics, chemistry, and other topics with their students. The laboratory is developing mathematics and science materials to be used by the National Urban Coalition in its Say YES to Youngsters program.

North Carolina Biotechnology Center, Research Triangle Park. This center is a state-funded, nonprofit corporation that promotes biotechnology research, business, and public awareness. Its biotechnology education project aims to increase the number of students receiving biotechnology education and to improve the quality of that education through a sustained effort to update a significant portion of the state's biology teachers about the science, applications, and issues of biotechnology. The center promotes teacher workshops, develops teaching materials, and provides teacher and student support services.

University of California Science and Health Education Partnership. Researchers at the University of California, San Francisco (UCSF) are paired with San Francisco Unified School District science teachers. UCSF researchers donate scientific and administrative equipment, update teachers on recent advances relevant to their curricula, provide technical advice, promote laboratory tours, and support an annual student science teaching contest. The partnership was founded in 1987 and is supported by grants from a variety of private foundations and most recently also from the U.S. Department of Education.

Student Enhancement:

Project WILD. This project is sponsored principally by state wildlife agencies and state departments of education as an interdisciplinary, supplementary environmental and conservation education program emphasizing wildlife. The project is based on the premise that young people and their teachers have a vital interest in learning about the earth. The goal of Project WILD is to assist learners of any age "to develop awareness, knowledge, skills, and commitment which will result in informed decisions, responsible behavior, and constructive actions . . . for wildlife and the environment upon which all life depends." Twenty-seven states sponsor the program. Activities have been developed for both elementary-school and secondary-school students.

APPENDIX F
PROGRAMS TARGETED TO WOMEN AND MEMBERS OF MINORITY GROUPS

The following are examples of programs intended to encourage women and members of minority groups to choose careers in the sciences. This compilation is not exhaustive, and much still needs to be done to recruit from groups that have not traditionally chosen science careers, but these examples show the range and types of resources currently available.

Carnegie Corporation of New York

The Carnegie Corporation of New York has sponsored numerous activities designed to improve the education of women and minority-group members. Grants have been awarded to a variety of organizations—the American Association for the Advancement of Science (AAAS), the National Urban Coalition, Fundacion Educativa Ana G. Mendez, Stanford University, the University of California, Berkeley, Arizona State University, the Science Museum of Connecticut, the Center for Applied Linguistics, the Center for Women Policy Studies, and the Institute for Educational Leadership—to promote equity and to prepare and motivate students to pursue and succeed in college mathematics and science-based courses. Some of the college programs are designed to mobilize community-based black and Hispanic parents' organizations in support of improved mathematics and science education for minority-group children. Others focus on middle- and high-school students. Still others are designed to build networks of colleges, community-based organizations, and professional associations to increase the access of women and minority-group members to higher education (Carnegie Corporation of New York, 1988). Funds for similar activities are also available from Apple Computer, the College Board, the Ford Foundation, the J.N. Pew Jr. Charitable Trust, and the National Aeronautics and Space Administration.

One recent example of an initiative to enhance exposure of minority-group children to science and mathematics is that of the National Urban Coalition (NUC). Recognizing that parents' fear of mathematics and science often affects students' interest in these subjects, NUC, under the sponsorship of the Carnegie Corporation of New York and others, has initiated a program that brings families together to participate in mathematics activities. The Say YES to a Youngster's Future program, now in its second year, targets teachers, children, and parents. Using mathematics materials developed by the Lawrence Hall of Science, teachers work with students and their parents. Family involvement is a vital part of the program, and children learn with their parents, siblings, or volunteer surrogate family members.

NUC is working with the Lawrence Hall of Science to develop science curricula for the Say YES project. An important component of both the mathematics and science projects is "cultural connections," historical information that links advances in science and mathematics to the ancestry of minority

137

groups, primarily blacks and Hispanics. For example, cultural connections include references to African theoretical mathematics, the African stone game, Egyptian and Mexican pyramids, the Mayan number system, and Mayan and Aztec calendars (National Urban Coalition, 1989).

The goal of NUC's project is to involve students in science and mathematics at an early age and, with the help and encouragement of family members, to sustain the interest. Although the NUC program is only in its second year, preliminary indications of the success of the program include observations that students are more motivated to become involved in extracurricular activities, that students show a greater level of participation in other classroom activities (e.g., spelling and geography bees), that there is greater student participation in science fairs, that test scores in mathematics have increased, and that teachers involved in the program have gone back to school to supplement their own knowledge of science and mathematics.

Other reservoirs of talent are black churches and black fraternal organizations. Project Linkages, a Carnegie-funded project at AAAS, has used black churches in efforts to retain black children in the science-engineering pipeline, and black fraternal organizations have a strong service orientation.

Howard Hughes Medical Institute

The Howard Hughes Medical Institute (HHMI) recognizes the need to attract minority-group students and women to careers in biomedical science. As one of its initiatives in undergraduate education, support is being given to colleges for high-school outreach programs. In 1988, 44 private 4-year colleges and historically black institutions were awarded grants to be used to prepare students for graduate education and for careers in research, teaching, or the practice of medicine. Ten of the 44 institutions that were awarded grants were historically black colleges or universities, and four were women's colleges. Among the projects to be supported are appointment of new faculty members and activities to develop faculty, curricular development and the acquisition of laboratory equipment, support of scientific research of faculty and students working in collaboration, academic development and scholarships for students in the sciences, and expansion and development of new linkages with teachers and students from high schools, junior high schools, and elementary schools (Howard Hughes Medical Institute, 1989).

In 1989, HHMI awarded grants focusing on colleges affiliated with doctorate-granting institutions. Support has been given for faculty, students, curricular development, and science outreach programs that strengthen the interaction of applicant institutions with other academic institutions, including not only 2- and 4-year colleges, but also elementary and secondary schools. Many of the applicant institutions have proposed initiatives to attract more minority-group members and women to careers in the biosciences through academic counseling, research and teaching assistantships, and sustained contact with senior research faculty in biology and related fields. Other programs are intended to offer summer institutes for biology teachers from schools with sizable minority-group populations. Outreach programs will focus on attracting minority-group students from neighboring community colleges to transfer to university science departments. Other programs will offer summer research assistantships for students in university laboratories (Howard Hughes Medical Institute, 1989).

Ford Foundation

The Ford Foundation sponsors programs developed specifically to attract minority-group students to the sciences. Ford Foundation doctoral and post-doctoral fellowships are designed to increase the presence of underrepresented minorities in the nation's college and university faculties. The Ford Foundation offers doctoral fellowships to members of six minority groups most severely underrepresented in the nation's Ph.D. population—Alaskan natives (Eskimo or Aleut), native American Indians, black Americans, Mexican Americans, native Pacific islanders (Polynesian or Micronesian), and Puerto Ricans. Awards are made for study in research-based doctoral programs in the behavioral and social sciences, humanities, engineering, mathematics, physical sciences, and biological sciences or for interdisciplinary programs comprising two or more eligible disciplines (Ford Foundation, 1988).

National Science Foundation

The National Science Foundation (NSF) minority-group graduate fellowship program was instituted to increase the number of practicing scientists and engineers who are members of ethnic minority groups that traditionally have been underrepresented in the nation's pool of scientists and engineers. The fellowships are awarded for study and research leading to master's or doctoral degrees in the mathematical, physical, biological, engineering, and social sciences and in the history and philosophy of science. Awards are also made for work toward a research-based Ph.D. in science education (National Science Foundation, 1988a,b).

NSF, through its Directorate for Science and Engineering Education, also sponsors activities to address the underrepresentation of women, minority-group members, and persons with disabilities. Support to set up regional centers for minority-group members and to develop prototype and model projects for women, minority-group members, and persons with disabilities is channeled through universities, colleges, and organizations with substantial enrollments of minority-group students in science and engineering and a demonstrated history of commitment to minority-group concerns (National Science Foundation, 1988c).

References

Carnegie Corporation of New York. 1988. The List of Grants and Appropriations of 1988. New York, N.Y.: Carnegie Corporation of New York.

Ford Foundation. 1988. Ford Foundation Predoctoral and Dissertation Fellowships for Minorities. Program Announcement 1989. Washington, D.C.: National Research Council.

Howard Hughes Medical Institute. 1989. Grants Program Policies and Awards 1988-1989. Rockville, Md.: Howard Hughes Medical Institute.

National Science Foundation. 1988a. 1989 Announcement for NSF Graduate Research Fellowships. Washington, D.C.: National Science Foundation.

National Science Foundation. 1988b. 1989 Announcement for NSF Minority Graduate Research Fellowships. Washington, D.C.: National Science Foundation.

National Science Foundation. 1988c. Guide to Programs. Fiscal Year 1989. Washington, D.C.: National Science Foundation.

National Urban Coalition. 1989. Miscellaneous "Say Yes" materials. Washington, D.C.

APPENDIX G
STRATEGIES FOR IMPLEMENTING
REPORT RECOMMENDATIONS

The committee cannot provide a complete blueprint or cost estimate for implementing the many recommendations in its report, but it describes here in rough terms the costs of implementing several components. We have chosen three examples and have explored them incompletely. The numbers used are based on assumptions and approximations gleaned from the literature and discussions with individuals involved in such activities.

Inservice Activities

The goal of new inservice activities is to reach as many science teachers as possible through useful and relevant programs and to sustain their involvement throughout their careers. We have tried to approximate the cost of engaging 27,000 biology teachers (75% of the estimated 37,000 biology teachers*) in various inservice activities over a number of years.

Each inservice activity is unique, and its costs pertain to numerous components, including the type of activity, instructors (level of expertise and number of staff involved), participant costs (various combinations of stipends, travel, and accommodations), supplies, equipment and other costs directly related to the activity, and indirect costs, which depend on the type of institution supporting the activity. Actual costs, therefore, might be considerably more or less than the estimates derived here.

We reviewed 20 National Science Foundation programs for biology-teacher enhancement on which relevant cost data were available (NSF, 1989). We calculated the average cost of inservice programs at $290 per teacher per day (in average 1989 dollars).

We assumed that 75% of biology teachers would participate in inservice activities. That percentage is considerably higher than the percentage of teachers who currently participate in inservice activities, but we expect that the development of new inservice activities will lead many more teachers to participate.

We also assumed that most new inservice activities would be set up to run for 2 workweeks (10 days) and be followed by two followup seminars—for a total of 12 days. (We also envision, however, that many other variations will develop.)

The following example illustrates the cost if 27,000 teachers participate in 12-day inservice activities that will scale up as new institutes offering summer inservice are developed over a 4-year period.

---

*Based on National Science Teachers Association and National Association of Biology Teachers estimates.

141

25% of teachers in first year:
$290 × 6,750 teachers × 12 days = $23.5 million
50% of teachers in second year:
$290 × 13,500 teachers × 12 days = $47.0 million
75% of teachers in third year:
$290 × 20,250 teachers × 12 days = $70.5 million
100% of teachers in fourth year:
$290 × 27,000 teachers × 12 days = $94.0 million
TOTAL: $235 million over 4 years for 75% of biol-
ogy teachers to attend inservice activities

In this case, the estimated cost at steady state for this program would be about $94 million per year in average 1989 dollars. It is important to note that the total approximate cost derived in this example is not considered to be a one-time expenditure. Teachers, both new and experienced, must participate in inservice activities annually throughout their careers. Funds must be provided through federal and other sources to develop, sustain, and evaluate these activities. If inservice activities were expanded in length and extended to all science teachers, the costs would, of course, increase proportionally.

If the goals described in this report are to be met, many new programs must be developed for teachers in the work force today, as well as for those who will enter the teaching profession in the future.

Mentors

The attrition rate among high-school science teachers (i.e., teachers not teaching science after 3 years) is 15% (Weiss, 1989, pp. 49-50, graph 37). On the basis of National Science Teachers Association and National Association of Biology Teachers estimates of nearly 37,000 biology teachers in the United States, approximately 5,550 new biology teachers are needed every 3 years. If, for example, one mentor teacher were assigned to two new biology teachers, there would be a need for 925 mentor teachers each year.

We made several assumptions about the costs of a mentor program:

• The average salary for a mentor teacher is $35,000.
• Mentor teachers will spend 20% of their time in mentoring activities (0.2 × $35,000 = $7,000).

The annual cost nationally would be approximately $6.5 million per year if each new teacher is to have a mentor for one year. This type of activity, however, is anticipated to occur in perpetuity, so budgets would need to be increased to reflect this recurrent cost. And, because science departments in most schools are too small to have two new biology teachers at once, perhaps a mentor teacher could be used by several schools or in some instances even by an entire school district. A number of variations on this theme are available; but the important element is to ensure that mentor teachers are available as resources to assist novice teachers in their critical first few years of teaching.

The preceding example does not address the costs of using mentor teachers to retrain "burned out" veteran teachers; to accomplish that goal, an even larger number of mentor teachers will be needed.

Fellowships for Teacher Education

The committee has recommended a competitive national fellowship to attract some of the most able biology or elementary-education majors to science teaching (see Chapter 6). In addition to attracting to the teaching profession bright and able students from liberal-arts colleges and universities that do not have traditional education programs, such a program might be of immense help in attracting members of minority groups that are underrepresented in the teaching force—people for whom the added expense of additional course work to become teachers would act as a disincentive.

As noted in the report, fellowships could be awarded both to students and to those who plan to make a career change. Participating institutions must have shown interest and imagination in addressing the kinds of changes that are required in preservice education. An example of such a commitment would be a program in which a "science-methods" course would truly integrate science and pedagogy.

Optimally, an institution that would participate in fellowship programs would:

- Be a research university with an accredited school of education.
- Have evidenced active collaboration between faculty of science departments and education departments.
- Have developed (or be planning to develop) distinct science-methods courses for future biology teachers.
- Emphasize approaches to science education that link content with pedagogical techniques.

Few institutions today would have all these characteristics. Criteria for selection of students must be developed. Gifted students who have expressed an interest in teaching biology would be particularly desirable. Some imagination will be required on the part of institutions, if they are to develop fellowship programs that will attract the brightest and ablest students to the teaching profession.

If, initially, 50 fellowships per year were awarded, the total annual cost would be $1 million plus the indirect costs, which would be different in each institution. If successful, the program should be expanded to service many more students with a proportionally higher cost. A fellowship would last for 1 year, at a cost per year of $20,000, including tuition. That cost might seem high, but one must consider that the types of students that such a program would try to attract could just as easily accept positions in other fields with much higher starting salaries. The small number of institutions that would be eligible to accept fellows, given the criteria set out above, might turn out to be the rate-limiting factor in attracting the ablest students. That is, not enough programs would be available.

Committee Members

Timothy H. Goldsmith (Chairman), a neurobiologist, is Professor of Biology at Yale University. He is a member of the National Research Council's Board on Biology.

Clifton Poodry (Vice Chairman), is Professor and Chairman, Department of Biology, University of California, Santa Cruz. A researcher in developmental biology, he has been active in the improvement of biology instruction for students on Indian reservations.

R. Stephen Berry is Professor of Chemistry at the University of Chicago with research interests in physical chemistry and in natural-resources allocation. He is a member of the National Academy of Sciences and holds a MacArthur Prize fellowship.

Ralph E. Christoffersen is Vice-President for Research, Smith Kline and French Laboratories. Trained as a chemist, he is a former president of Colorado State University.

Jane Butler Kahle is Condit Professor of Science Education and Professor of Zoology at Miami University, Ohio, and former president of the National Association of Biology Teachers. She has published numerous articles on teaching secondary-school science and on women and minority groups in science education.

Marc W. Kirschner is Professor of Biochemistry at the University of California, San Francisco, and does research on the regulation of cell growth and cell division. He is a member of the National Academy of Sciences.

John A. Moore, Professor of Biology, emeritus, University of California, Riverside, led the team that developed the Yellow version biology text of the Biological Sciences Curriculum Study in the 1960s and early 1970s. He is the current Director, Science as a Way of Knowing project, American Society of Zoologists, and a member of the National Academy of Sciences.

Donna Oliver is the 1987 National Teacher of the Year, designated by the Council of Chief State School Officers, Good Housekeeping, and Encyclopaedia Britannica. She is Associate Professor of Education at Bennett College, Greensboro, North Carolina.

Jonathan Piel, current editor-in-chief of *Scientific American*, has extensive experience in science journalism as writer and editor. He is a graduate of Harvard College.

James T. Robinson is former Executive Director, Curriculum and Evaluation, Boulder (Colorado) School District. He served as a staff officer for the Biological Sciences Curriculum Study and was on the faculty of Teachers College, Columbia University.

Jane Sisk teaches biology at Calloway County High School, Murray, Kentucky. She was recognized as an Outstanding Biology Teacher by the National Association of Biology Teachers in 1983 and was 1984 recipient of a Presidential Award for Excellence in Science Teaching. She was also the 1987 recipient of the Christa McAuliffe Fellowship.

Wilma Toney teaches biology at Manchester High School, Manchester, Connecticut. She formerly taught in the primary schools of the District of Columbia.

Daniel B. Walker is Associate Professor of Biology and Science Education at San Jose State University, California. His research is in plant development.

SPECIAL ADVISERS

Paul DeHart Hurd is Professor of Science Education, emeritus, at Stanford University. Long a leader in science-curriculum development, he is a member of the human-biology program under development at Stanford. He has been associated with the Biological Sciences Curriculum Study since its origin.

John Harte holds a joint professorship in the Energy and Resources Group and the Department of Plant and Soil Biology at the University of California, Berkeley. He is also a senior faculty researcher at the Lawrence Berkeley Laboratory and a senior investigator at the Rocky Mountain Biological Laboratory.

Index

147